KT-142-414

WITHDRAWN

APPLIED
ROBOTICS

LIBRARY
RUGBY COLLEGE

RUGBY COLLEGE OF FE

* 0 0 6 3 0 8 8 7 *

WITHDRAWN

LIBRARY
RUGBY COLLEGE

Dawson £22.50

LIBRARY RUGBY COLLEGE	
ACCESS No.	
CLASS No.	
DATE	

LIBRARY
RUGBY COLLEGE

APPLIED
ROBOTICS

EDWIN WISE

PROMPT®
PUBLICATIONS

```
┌─────────────────────────────────┐
│           LIBRARY               │
│        RUGBY COLLEGE            │
├──────────┬──────────────────────┤
│ ACCESS   │                      │
│ No.      │   630887             │
├──────────┼──────────────────────┤
│ CLASS  ✓ │   629.892            │
│ No.      │                      │
├──────────┼──────────────────────┤
│ DATE     │   - 8 OCT 2001       │
└──────────┴──────────────────────┘
```

©1999 by Howard W. Sams

PROMPT© Publications is an imprint of Sams Technical Publishing, 5436 W. 78th Street, Indianapolis, IN 46268.

All rights reserved. No part of this book shall be reproduced, stored in a retrieval system, or transmitted by any means, electronic, mechanical, photocopying, recording, or otherwise, without written permission from the publisher. No patent liability is assumed with respect to the use of the information contained herein. While every precaution has been taken in the preparation of this book, the author, the publisher or seller assumes no responsibility for errors or omissions. Neither is any liability assumed for damages resulting from the use of information contained herein.

International Standard Book Number: 0-7906-1184-8
Library of Congress Catalog Card Number: 98-068716

Acquisitions Editor: Loretta Yates
Editor: J.B. Hall
Assistant Editor: Pat Brady
Typesetting: J.B. Hall
Indexing: J.B. Hall
Cover Design: Christy Pierce
Proofreader: Patika Bush
Graphics Conversion: J.B. Hall
Illustrations and Other Materials: Courtesy of the Author

Trademark Acknowledgments:
All product illustrations, product names and logos are trademarks of their respective manufacturers. All terms in this book that are known or suspected to be trademarks or services have been appropriately capitalized. PROMPT® Publications and Sams Technical Publishing cannot attest to the accuracy of this information. Use of an illustration, term or logo in this book should not be regarded as affecting the validity of any trademark or service mark.

PRINTED IN THE UNITED STATES OF AMERICA

9 8 7 6 5 4 3

LIBRARY
RUGBY COLLEGE

Contents

Chapter 1
Introduction To Robotics 1

Chapter 2
Mechanical Platforms 7

Chapter 3
Power Supplies 23

Chapter 4

Electronic Control 35

Chapter 5

Adding Sense 49

Chapter 6
It's Alive! Simple Robot Behaviors 65

Chapter 7
Microcontroller 71

Chapter 8

MCU Senses........................... 99

Chapter 9

Igor, Fetch Me Some Brains 121

Chapter 10
Fuzbol 141

Chapter 11
More Sense 173

Chapter 12
R/C Servos 191

Chapter 13
Pneumatics 197

Appendix A: Fuzbol Language Reference 213

Appendix B: Sample Programs 235

Appendix C: Conversions and Code References 285

Suppliers 289

Preface

I would like to thank my father for inspiring me to be a Mad Scientist. Loretta Yates, I must thank you for taking a chance on me for this book. My wife Melissa – well, she stood behind me no matter how mad the projects. James and Jessica, thanks for the parts!

I am writing this book on robots because I enjoy solving problems – and robots present more problems to the inventor than any other system. Robotic systems cover all the bases – mechanical design, sensory systems, electronic control, and computer software intelligence. This book addresses each of these issues and the result is a small, mobile, self-controlled robot.

This book gives a project-oriented introduction to the field of robotics. It guides the beginner through the challenges of building a working robot and it may provide new ideas and techniques for the advanced builder. No single project in this book is very difficult; in fact, most of them are extremely simple. This reflects my philosophy towards hard projects. Inside of every complex system is a set of simple systems waiting to get out. If you try to build a huge, complex system all at once you quickly become overwhelmed with details and problems. Instead, build and understand the many simple systems, each of which is fairly easy. How do you eat an elephant? One bite at a time.

Wherever possible the projects use parts readily available from Radio Shack, though many MCU projects use parts only available through mail-order.

Updates, errata, supporting information, and selected project kits may be found at http://www.simreal.com. The author can be contacted at ewise@simreal.com.

CHAPTER 1
Introduction To Robotics

Robot — "A reprogrammable, multifunctional manipulator designed to move material, parts, tools, or specialized devices through various programmed motions for the performance of a variety of tasks." *Robot Institute of America, 1979*

Robots?

Since mankind's earliest time, life and death have fascinated us… as has the possibility of creating new life through our own efforts. In a way, the early tales of the Golem (a clay creature brought to life through certain spells and incantations) mirrors both the creation of Adam brought to life from the dust of the earth and our own desire to build robots.

Of course, we didn't have the name "robot" for this creature until Karel Capec wrote his classic *Rossum's Universal Robots* in 1921. His machines, as in so many stories since, turned against their creators.

Capec's "robots" were much like Dr. Frankenstein's monster, creatures created by chemical and biological rather than mechanical methods. The mechanical robots of popular culture are not much different from these biological creations. They look, act, and emote much like the humans they are designed to emulate. However, that isn't what robots are typically like.

The average robot today is used in the factory or assembly line and, at its anthropomorphic best, consists of an arm and a gripper, vaguely reminding us of a dismembered arm in service to an endless row of parts marching before it on the conveyor. As hobbyists, we are not building Frankenstein's patchwork creature from spare parts dug up in the middle of the night, nor are we looking to re-create the high-precision, highly durable equipment used on the factory floor. We are here to learn, explore, and play. For our purposes a robot isn't a "specialized device with various programmed motions for the performance of a variety of tasks." It is a machine that senses its environment and reacts to it independent of human intervention.

The main thread through this book is a small, wheeled robot that serves as the test platform for a variety of experiments in sensing the environment and making intelligent choices in response to it. The "intelligence" of this robot is below any measurable limit, yet through our efforts it will transform from a pile of inert parts into an entertaining "pet."

At a very basic level, a robot consists of:

1. A mechanical device, such as a wheeled platform, arm, or other construction capable of interacting with its environment.

2. Sensors on or around the device that are able to sense the environment and give useful feedback to the device.

3. Systems that process sensory input in the context of the device's current situation and instruct the device to perform actions in response.

Of course, this definition covers such mundane things as your home thermostat. We should have more fun than that!

Tools and Equipment

I know a man who built a baby's cradle using nothing more than a pile of wood and his pocketknife. It is beautiful and I can't imagine how he did it. Given the choice, I would use a different set of tools. While it is *possible* to create all of the projects in this book with a minimal set of tools and test equipment, the more tools you have the easier all of this becomes.

The minimal tool set would include an electric drill and bits, small screwdrivers, small pliers, a wire cutter, wire stripper, soldering iron, and a multimeter. A better set of tools adds such things as an oscilloscope, higher-quality variable-temperature soldering iron, tweezers, and a third-hand vise. Whenever buying tools, get the best you can afford and you will never regret it.

Almost all problems can be defined using the black-box model of input, processing, and output. Without tools you have to guess at what is happening. Good tools give you good information. They take the top off of the black box and show you what is really going on. Since a project rarely works the first time, anything that helps with the debugging process pays for itself in time and frustration. When building projects, especially the microcontroller-based ones, a good sense of paranoia can be very handy. Trust nothing, rely only on what can be seen on the scope or through the readouts. Understand each chip in the project and know what to expect from each pin. Read the manufacturer's data sheets and try your best to understand them. And remember – nothing is ever entirely free of errors, so stay alert.

For electronics projects I recommend using something like the Global Specialties Proto-Board for all of the prototyping and testing *(Figure 1-1)*. You can get this type of prototyping board at any electronics supply store, and they come in many different sizes and configurations. The one used in this book is a Global Specialties UBS-100 which has a connection pattern as shown in *Figure 1-2*. I like this board because of its two separate power distribution strips (one set on each side). If you use one with a different number of power strips, you will need to modify the board layouts for some of the projects. I always take a permanent felt-tip pen (such as a Sharpie®) and mark the holes on the power strips, red for power and black for ground.

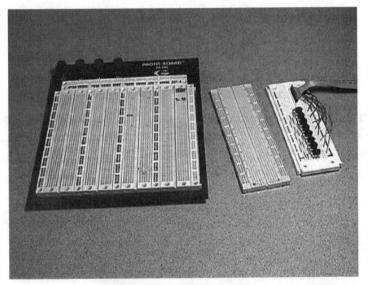

Figure 1-1. Prototyping boards.

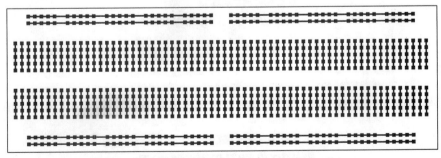

Figure 1-2. Prototyping board connection patterns.

A good tool in any experimenter's toolkit is a pad of grid paper. Take plenty of notes and keep them forever.

Finally, a tool that is easily overlooked is the online or mail order catalog. Many parts can be had quickly through the mail, purchased from catalogs or off the Internet. Before you begin the projects in this book, I recommend that you find the on line catalogs or order the paper catalogs from these companies.

Radio Shack

Radio Shack outlets can be found everywhere in the US. Though more of a consumer-product company, they carry basic supplies for the experimenter, and their series of workbooks by Forrest M. Mims III are excellent for the beginner.

You can shop at Radio Shack outlets, or phone in your order. They do not have online shopping at this time.

1-800-THE-SHACK (1-800-843-7812)
http://www.radioshack.com

Tech America

Tech America is a new service with catalogs sold through Radio Shack outlets. Where Radio Shack has just a few parts, Tech America is all parts and test equipment. Tech America let's you order by phone and they have online ordering as well.

1-800-877-0072
http://www.techam.com

Digi-Key

Digi-Key has an extensive catalog and excellent service. You can place orders both online and by phone.

1-800-344-4539
http://www.digikey.com

Mouser

Mouser is another extensive electronics catalog with phone and online ordering.

1-800-346-6873
http://www.mouser.com

Small Parts, Inc.

On the mechanical side of robotics, Small Parts carries an amazing variety of, well, small parts.

1-800-220-4242
http://www.smallparts.com

The Robot Store

Though a small catalog, the Robot Store caters specifically to the robotics enthusiast and is worth checking out.

1-800-374-5764
http://www.robotstore.com

CHAPTER 2
Mechanical Platforms

Without some type of "body," a robot isn't a robot at all but some form of artificial intelligence or AI system. So let's build a body.

There are many different bodies that can be called a robot. For example, articulated arms used in welding and painting, gantry and conveyor systems for moving parts in factories, and giant earth moving machines deep inside mines. One of the most interesting things about building a robot is the creation and observation of its behavior, and the simplest behavior is locomotion.

There are many different forms of locomotion in nature; insects with their many legs (*too* many, if you ask me), mammals with their (mostly) four legs, birds with two, and worms and snakes without any legs at all. Legs are complicated mechanical systems, so we're not going to do legs. Except for microscopic organisms and their occasional flagella, nature doesn't make much use of rotary motion. Wheels are the simplest way for *hobbyists* to make something move, so that is what we will use.

Wheels by themselves aren't very interesting. You need to provide some force to make them turn under command. Radio control cars and airplanes have a variety of gas-powered engines. Pneumatic tools use air to spin their shafts. We are going to use electricity, because it is simple.

What size should the robot be? There are robots so small that they fit into a thimble, and MIT has a nice team of robots that fit into a 1" cube. On the other side of the scale, there are robot construction machines used in mines and other dangerous areas that wouldn't fit into the average garage. Extremes in size, both large and small, present interesting challenges in construction, power supply, and control. As an introduction, work in an easy size, roughly 6" square, a natural size for the inexpensive components we are working with.

Types of Electric Motors

There is a bewildering array of electric motors on the market. Let's touch briefly on a few basic types. A detailed (and animated) tour of several types of motors can be found at http://mot-sps.com/motor/mtrtutorial/mtr.html. Also check out the Aveox motor primer at http://www.aveox.com/primer.html.

First we will define some terms and values associated with motors. The strength of a motor is measured in several ways. Torque is the measure of how hard the motor can push. Torque varies inversely with speed, from the maximum push when the motor is stalled, down to the minimum torque at the motor's maximum speed. The most efficient power output from an electric motor is at its mid-speed. The push of the motor is measured in ounce-inches or foot-pounds, which describes how much force the motor exerts at a particular distance from its axis center.

Another measure of a motor's strength is in Watts of electricity consumed. The Watt rating of a motor can be calculated from its current draw, voltage, and efficiency ratings:

$$W = V \times I \times \text{efficiency}$$

Or, Watts can be calculated from the motor's torque at a particular speed:

$$W = (\text{ounce-inches} \times \text{RPM}) / 1352$$

Finally, Horsepower and Watts are related by a simple power factor:

$$1 \text{ HP} = 746 \text{ Watts (at 100\% efficiency)}$$

The efficiency rating of a motor describes how much of the electricity consumed is converted to mechanical energy.

DC Motors

Your garden-variety PMDC (permanent magnet, direct current) motor is easy to use. Attach power to the two leads, and away it goes! PMDC motors use an arrangement of fixed and electro-magnets (stator and rotor) and switches (commutator) to create motion through a spinning magnetic field. Small PMDC motors are used in most motorized toys, and most of these seem to be made by Mabuchi.

AC Motors

An AC Motor doesn't use a commutator to spin the magnetic field. Instead, they cycle the power at the input leads to continuously move the field. This is also the concept behind brushless DC motors.

Stepper Motor

A stepper motor is like the brushless DC or AC motor. They move the rotor by applying power to different magnets in the motor in sequence (stepped). Steppers are designed for fine control and will not only spin on command, but can spin at any number of steps per second (up to their maximum speed). Steppers can also hold their position and resist turning, which most other motors can't do.

Servo Motors

The previous motors are open loop. Given a signal, they perform their action to the best of their ability. A servomotor is a closed loop device. Given a signal, it adjusts itself until it matches the signal. If conditions change (such as the external force on the motor shifts), the motor changes to match them.

Servos are used in radio control model airplanes and cars. They are simple DC motors with gearing and a feedback control system. Model servomotors are designed to oscillate (move back and forth) rather than rotate, though it is possible to physically modify them to serve as drive motors.

Using PMDC Motors

Off the shelf, most motors have their best efficiency at a high speed and low torque. Attach a couple of wheels to your basic toy DC motor and power it up. If it moves at all, it will zip along too quickly to be useful. Almost every application for a PMDC motor requires that it be slowed down.

A side effect of mechanically reducing the speed of a motor is that it also increases the strength of its output. As a general rule, energy can neither be created nor destroyed. Ignoring energy wasted through friction, when a system reduces one part of its equation by some amount, another part of the equation must increase by the same proportion. In plain terms, if you reduce the velocity of a motor through a reduction system, you also increase its strength. This is the principle behind levers, gears, and pulleys.

Gears and Chains

Gears and chains provide a strong and accurate way to transmit rotary motion from one place to another, possibly changing it along the way. Gears and chains are found on bicycles, motorcycles, and in other high-strength applications. Because of the way they mesh together with the gear, chains don't slip. Their metal construction also makes them very strong. If the gears are not well aligned, however, the chains can walk off of the gears.

The speed change between two gears depends on the number of teeth on each gear. When the powered gear goes through a full rotation, it will pull the chain by the number of teeth on that gear. The chain then pulls the other gear by that same number of teeth, which can be more or less than a full rotation.

For example, let's examine a powered gear, A, with 10 teeth and a slave gear, B, with 40 teeth, as shown in *Figure 2-1*. Every rotation of gear A moves the chain 10 teeth worth of motion. Since B needs 40 teeth worth of chain motion to make a full circle, it only moves 1/4 of a circle, reducing the speed at that end to 0.25 and increasing the strength by 4.

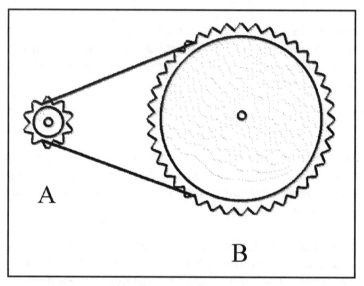

A

B

Figure 2-1. Gear and chain.

If A is rotating at 10,000 revolutions per minute:
$$B_{rpm} = A_{rpm} \times (A_{teeth} / B_{teeth})$$
$$B_{rpm} = 10,000 \times (10 / 40)$$
$$B_{rpm} = 2,500$$

One problem with gear chains (and, later, pulleys and belts) is that if you want make a huge reduction in speed, you have to string a bunch of these gears together, and this can become awkward and inefficient. As a rule of thumb, no single stage of reduction should be more than about 5 to 1, to prevent undue stress on the system.

Pulleys and Belts

Pulleys and belts are used in much the same way as gears and chains. Pulleys are everywhere – in your car, refrigerator, and clothes dryer. Pulleys are wheels with a groove around the edge, and belts are the rubber loops that fit in that groove. For high-torque or high-accuracy situations, there are toothed belts to avoid slipping (such as the timing belts in your car). Pulleys and belts are inexpensive, easy to use, and work for many situations. These systems are also less likely to jump off of their track than chains.

The amount of speed reduction you get between two pulleys depends on their relative sizes, the same as in gears. Instead of counting the teeth around the edge, you use the circumference of the pulley.

For example, examine the system where pulley A is 1" in radius and the output pulley B is 4" radius as shown in *Figure 2-2*. Let's assume that A spins at 10,000 rotations per minute.

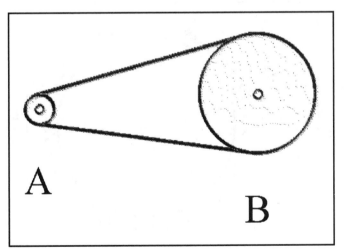

Figure 2-2. Pulley and belt.

$$A_{circum} = 2 \times \pi \times A_{radius}$$
$$A_{circum} = 6.283$$

$$B_{circum} = 2 \times \pi \times B_{radius}$$
$$B_{circum} = 25.133$$

$$B_{rpm} = A_{rpm} \times (A_{circum} / B_{circum})$$
$$B_{rpm} = 10,000 \times (6.283 / 25.133)$$
$$B_{rpm} = 2,500 \text{ RPM}$$

Since the circumference is based on the radius, you can simply take that ratio:

$$B_{rpm} = A_{rpm} \times (A_{radius} / B_{radius})$$
$$B_{rpm} = 10,000 \times (1 / 4)$$
$$B_{rpm} = 2,500 \text{ RPM}$$

Gearboxes

Gearboxes are everywhere – the transmission in your car, the timing mechanism in a grandfather clock, and the paper-feed of your printer.

A Gearbox operates on the same principle as the gear and chain, without the chain. Gearboxes require closer tolerances since, instead of using a large loose chain to transfer force and adjust for misalignments, the gears mesh directly as shown in *Figure 2-3.*

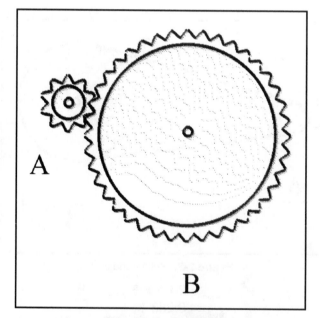

Figure 2-3. Gearbox.

Unless you have a decent machine shop (or a good LEGO™ set), you need to buy pre-made gearboxes. In that case you can pick from a list what type of reduction you want. There are even gearboxes that fit directly onto a motor, sold together as a unit called a gearhead motor.

Cheating

With a decently equipped workshop, a handful of gears, some metal or plastic, and a dose of profanity, you can make a gearbox for a motor. Attach it to a platform, slap some wheels on it, and you are ready to start giving the robot behaviors. Or you can look around the toy and hobby stores and see what shortcuts are available.

Perhaps the simplest shortcut is to buy a pre-built wheeled platform. Specialty robot hobbyist catalogs have a variety of nice robot bases. Buying a platform pre-made takes some of the fun out of it, plus it can be an expensive option.

The toy and hobby stores are great sources of inspiration. I can spend hours at our local hobby shop looking over tiny parts, thinking about their possible uses. Aha! There, in the middle of the toy store, is a huge pile of radio controlled cars. Some of them, especially the "wired-remote" cars for the little kids, are a good size and a good price. They may not do us much good, though. Not because of anything obvious like price, size, or speed, but because of the way that they are steered.

There are two fundamental ways to change the direction of a wheeled vehicle (that is, to steer it). One is "car" steering, what you see in 90% of the R/C cars (see *Figure 2-4*). The rear drive wheels move the vehicle forward and backward, and the front steering wheels change orientation left and right to make the car point in different directions. This is great for roads, but in a house where there are tight corners, dead ends, and other traps, the robot may need a tighter steering radius than any car.

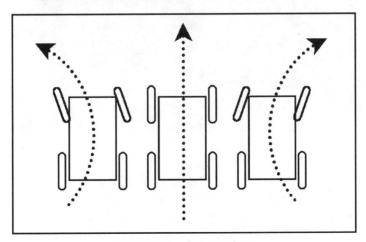

Figure 2-4. Car steering.

The other steering method is "tank"(differential) steering, as shown in *Figure 2-5*. In tank steering, there are only drive wheels. When both drive wheels are going forward the vehicle moves ahead. When they both go backwards it is in reverse. However, if the drive wheels are going at different speeds, or in different directions, the platform turns. Sometimes you can find off-the-shelf toys with tank steering, like the one in *Figure 2-6*, and they can work nicely. Buy the cheapest one you can find that meets your size needs, and remove everything that isn't necessary (including the electronics, if any).

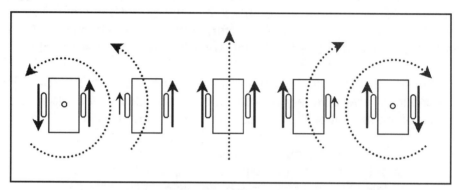

Figure 2-5. Tank steering.

Figure 2-6. Off-the-shelf tractor base.

Where the wheels are mounted and the shape of the platform itself makes a big difference in how the robot moves. The simplest form of all is a circular robot with the drive wheels across the center and castors at the front and/or back for stability. Any other configuration adds corners to snag and a more complicated steering geometry.

A good source of gears and motors can be found as add-ons for almost every modular construction system at the toy store, such as LEGO™, K'Nex™, Meccano™, and Fischertechnik™. These can be expensive, but you could always buy them for the kids and then use them later, when the kids are "bored" with them.

It is also possible to convert radio control servomotors so they act like gearhead motors. This conversion requires a bit of disassembly, removal of the position-sensing potentiometer, and some other circuit changes that have been covered in great detail by other authors. In short, carefully disassemble the servomotor, cut the "stop" off of the large gear, and replace the potentiometer with a balanced pair of resistors. Tamiya makes some nice gearboxes for the hobbyist, such as the one shown in *Figure 2-7*. They come in a range of gear ratios and are perfect for building a robotic platform. If you can't find them at the local hobby shop, go to Tamiya's web page for a list of dealers in your area.

Figure 2-7. Tamiya gearbox and motor.

Project 2-1: Building a Small Wheeled Platform

As with all projects, read through the entire set of instructions before you start working on the project. You'll save lots of time and money.

Qty.	Part	Description
2	Gearbox	Tamiya wormgear motor, or equivalent
2	Wheel, 2.5" dia.	To fit the Tamiya gearboxes
1	Wheel, 1" dia.	To fit the stiff wire, for the tail-dragger
	Wire, stiff	For the tail-dragger wheel
1	Perforated Board, 6" x 8"	Cut to size for the robot's body
1	Prototyping Board	For electronics experimentation.
	Bolts, 4-40, with nuts	Or similar, 1/2" to 1" size, to hold it all together
	Spacers	Or similar, to keep things apart
3	Shaft Collars	To hold the tail-dragger wheel and wire together

Table 2-1. Project 2-1 parts list.

The platform uses two Tamiya high power worm-gear motors to provide drive power in conjunction with a pair of model airplane wheels. Each Tamiya unit has a small DC motor, a gearbox, and a variety of mounting hardware.

After assembling the gearbox according to the instructions (or otherwise finding an operating motor and gearbox), you may need to modify the wheels to fit properly. The Tamiya gearbox is designed to attach to the (included) plastic arms and disks, but not to a wheel.

To mount the wheel you first have to drill two holes at the appropriate location to bolt it onto the mounting disk *(Figure 2-8)*. Attach the disk to the gearbox *(Figure 2-9)* and then mount the wheels, bolting them onto the disks *(Figure 2-10)*. You may need to cut the drive shaft on one side of the motor, so it doesn't interfere with the motor on the other side of the robot.

The platform itself is made from a piece of perforated board, the kind used in electronics prototyping. Take a 4.5" by 6.5" board (or cut down a larger board) and lay the gearboxes and prototyping board on it to get a sense of size. To keep from working yourself into a corner, look ahead to Project 4-4 and Project 5-2 where additional hardware is attached to the board. See *Figure 2-11* for the layout used for the projects in this book.

Figure 2-8. Wheel with new mounting holes.

Figure 2-9. Mounting disc attached to gearbox.

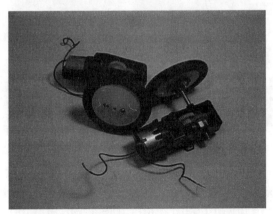

Figure 2-10. Wheels attached to gearboxes.

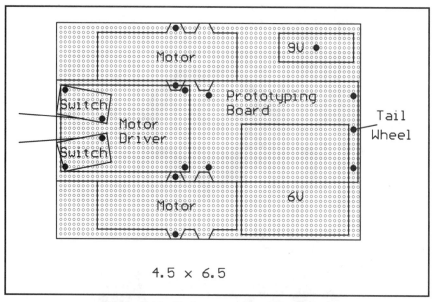

Figure 2-11. Perforated board parts layout.

If you can only find a larger board, cut it down to 4.5" by 6.5" or other suitable dimension. I would normally use a small jigsaw or bandsaw to cut the board, but what I had on hand were my trusty wire cutters. Amazingly enough, the perforated board cut very cleanly with these. Cut away small strips at a time (*Figure 2-12*).

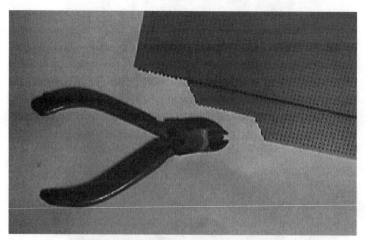

Figure 2-12. Cutting the perforated board.

Place the two gearboxes so that their axles are opposite each other, leaving room for battery packs, the motor driver board from Project 4-5, and the feeler switches from Project 5-2. Mark the holes on the perforated board where the gearboxes bolt down.

You can place the wheels at the exact center of the platform, or you can place them more to one end. Later you will add a third wheel and place weight (the batteries) between it and the drive wheels to make the platform stable.

Lay the prototyping board along the center and mark its holes. Drill all of the holes to a size where the bolts fit nicely. It's better to make them a little large than too snug.

Bolt the gearboxes and the prototyping board onto the platform. You may want to use standoffs (little metal or plastic tubes) to raise the prototyping board up above the gearbox's bolts. If you raise the prototyping board, you need to reinforce the paper on the bottom of it. The board's press-in connectors are held in place by a sheet of paper stuck to the back of the board. When this is raised off of a hard surface, it can come unstuck at the ends and allow the connectors to fall out. A little duct tape can reinforce this nicely.

If you use the remnants of a toy as a mobile platform, you only need to drill holes in the toy and mount the prototyping board on it, as shown in *Figure 2-13,* since the rest of the drive train is already in place.

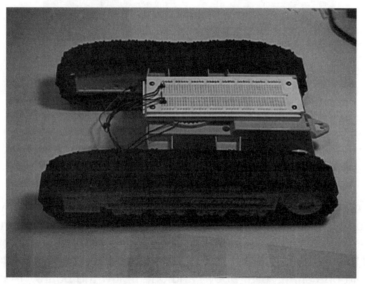

Figure 2-13. Tractor base with prototyping board.

The two-wheeled platform is a bit tippy right now. To balance it you need to add a third "tail-dragger" wheel. This coaster wheel is pulled behind the robot and makes it a stable three-wheeled platform. Small wheels designed for model airplanes, and sometimes the entire mounting hardware, can be found at the local hobby shop.

Start building the third wheel by finding either a small wheel, or the stiff wire. These should each fit the other. A good source of steel and brass wires in many gauges can be found at the local welding supply house. Copper-clad steel wire is used for welding, and brass wire is used for brazing. You can also find "piano" wire at most hobby shops, but this can be too brittle for many bending operations. The advantage of the hobby shop is that you can buy just one wire, where the welding supply may want you to walk off with ten pounds or more.

Once you have a wheel and some wire, bend the wire into a suspension as shown in *Figure 2-14*. This suspension needs a short horizontal section where the wheel attaches, a vertical section that goes up the radius of the wheel (and then a bit), another short horizontal section that reaches back to the centerline of the wheel, and then the final vertical segment to reach up and into the robot platform.

Figure 2-14. Third wheel.

The wheel is held onto the wire, and the wire is held onto the platform, by small shaft collars and "grub screws." These little collars slip onto the wire and the grub screws in to pinch the wire tightly within. Attach the wheel to the wire using one of these.
You may need to enlarge a hole in the perforated board for the wire. Attach the wire to the board with one shaft collar above and one below the board. Placing a small washer on either side of the perforated board helps the tail-wheel to turn easily. Later, when the battery packs are attached to the platform, be sure to leave enough clearance around this wheel's mounting collars and suspension wire so it can pivot freely. That's it! (*Figure 2-15*) It doesn't *do* anything, but it is the skeleton for all of the other projects in this book.

Figure 2-15. Third wheel attached to the platform.

CHAPTER 3
Power Supplies

Dr. Frankenstein harnessed the powers of nature to bring his creation alive, drawing the power of lightning down from the heavens and channeling it through his arcane mechanisms to re-spark whatever vital essence still resided within the assembled flesh.

Sounds like fun, doesn't it? In most of the projects in this book, we also harness the power of electricity to make things move, but electricity isn't the only option.

All forms of energy are related and they can be converted between one another with an associated loss of energy relative to the efficiency of the conversion. Mechanical energy (motors, solenoids) is converted to electrical (dams and windmill's spinning generators), and electrical into chemical (charging batteries or splitting water into hydrogen and oxygen). Chemical energy converts into mechanical (gasoline exploding in engines) and electrical (batteries, fuel cells), and so forth. These conversions are put to good use in the sensors added to the robot later.

Most portable power is based on chemical sources – gasoline, batteries, and fuel cells. Of all the various forms of power available, batteries are by far the simplest, least expensive, and most common. Solar cells are another option for portable power, and are the heart of BEAM Robotics (a particular branch of robot building that emphasizes natural designs). However, solar cells don't provide enough power to drive the systems in this book.

Batteries

There are two types of battery: Primary batteries are used once and then discarded; secondary batteries operate from a (mostly) reversible chemical reaction and can be recharged many times.

Primary batteries have higher energy density (they give more sizzle for their size) and a lower self-discharge rate (they last longer on the shelf).

Secondary (rechargeable) batteries don't have as much energy as primary batteries, but they can be recharged up to a thousand times depending on their chemistry and environment. Some rechargeable batteries show a "memory effect" where, if the battery is consistently discharged to a particular level it slowly loses the ability to discharge any deeper, effectively reducing its capacity. This can be avoided by simply draining those batteries completely each time they are used.

Primary batteries are discarded when they are used up. You discard secondary batteries when their life cycle is reduced by half. At the first use of a rechargeable battery it may last for 4 hours of continuous operation in an application. Once that same battery lasts for only 2 hours, toss it.

Choosing a battery for a project means deciding among the literally hundreds of types and styles of batteries available. Batteries are categorized by their chemistry and size, and rated by their voltage and capacity. The voltage of a battery is determined by the chemistry of the cell, and the capacity by both the chemistry and size. Battery sizes are shown in *Table 3-1*.

Size	NEDA	IEC	Description
AAA	24A	LR03	Smallest of the common sizes.
AA	15A	LR6	Most popular small battery, typically used in packs of 2 or 4.
C	14A	LR14	Small flashlight battery, large toys.
D	13A	LR20	Largest common battery, good for hungry applications with lots of space.
9v	1604A	6LR61	Rectangular, with clip-on connector.

Table 3-1. Battery sizes.

The batteries on the shelf at the grocery store come in five common sizes. Other sizes range from the very small button-cells found in watches and hearing aids, to the large batteries in cars and wheelchairs.

Zinc-Carbon

Zinc-carbon batteries used to be the most popular household primary batteries, until Alkaline came along and took the spotlight. Zinc-Carbon batteries are less expensive and are best used in medium to low-power applications. They have about half the capacity of alkaline batteries.

Alkaline

The alkaline primary battery has become the most common battery in use. It is reliable, supplies more power than any other household battery, has a long shelf life, and can be found everywhere. It is recommended for higher-power applications, but can be used anywhere. *Table 3-2* lists the basic attributes for common alkaline battery sizes.

	AAA	AA	C	D	9v
V (nominal)	1.5	1.5	1.5	1.5	9
V (discharged)	.8	.8	.8	.8	4.8
Capacity (mAH)	1,000...1,200	2,100...2,800	7,000...7,800	13,000-15,000	550...580
Drain (mA)	15	28	60	115	12

Table 3-2.

V (nominal) indicates the optimum voltage of a fresh battery.

V (discharged) indicates how low the voltage goes before the battery is considered dead. Notice how low the voltage goes before the battery is finished – almost half of the nominal voltage. Different battery types approach this discharge voltage over time along different curves... if you need to know the details, each manufacturer has complete specifications for their products.

Capacity is how much power the battery can provide, and is specified in Amp-Hours (AH) or milli-Amp Hours (mAH). A milli-Amp is 1/1,000 of an Amp (or, 1,000mA equals 1A). A milli-Amp Hour is what you get when you multiply the mA drain on the battery by the number of hours you drain it. For example, a 2,500 mAH cell could (in theory) put out 2.5A in one hour of heavy use, or 100 mA over 25 hours of medium use. In practice, there is a maximum rate of drain you can apply to a battery without shortening its life.

Drain (mA) This is the drain rate the batteries were tested at to determine their capacity. Higher drains may yield lower overall capacities.

Rechargeable Alkaline

These are fairly new on the market. They have the longest shelf life of the rechargeable batteries. Though the rechargeable alkalines don't have a discharge memory, their performance does degrade with each recharge.

Nicad

NiCd, Nickel-Cadmium. These are the most common rechargeable batteries, and like alkaline batteries, they are quite rugged. They can be recharged about 750 times, depending on their use, memory, and environment. Extreme environments are hard on batteries – heat, cold, very high or very low humidity. They all take their toll. *Table 3-3* lists the basic attributes for common Nicad battery sizes.

	AAA	AA	C	D	9v
V (nominal)	1.2	1.2	1.2	1.2	7.2
Capacity (mAH)	180-220	500-650	1,200-2,500	4,000-5,000	80
Drain (mA)	15	50	225	450	

Table 3-3.

Nicads have the strongest memory effect and you should make sure you drain them fully before recharging.

Note that Nicad batteries run at a lower voltage than alkalines – 1.2 volts instead of 1.5 volts. This can make a big difference depending on the application, since a pack of 4 Nicads reaches only 4.8 volts (at best), and many electronic applications need at least 5 volts. Nicads also have much lower capacity than alkalines, as bad as 1/5th and as good as 1/3rd the mAH rating. On the plus side, you can safely apply higher loads to them.

Nickel

NiMH, Nickel Metal Hydride. Not the most common package on the shelf, and not always found in the standard sizes. NiMH batteries can have up to twice the capacity of Nicads and show less of a memory effect. They have a 1.2 volt standard, can be recharged about 400 times, and have a higher rate of self-discharge than Nicads.

Lithium

Lithium batteries are expensive and are used for specialty electronics (such as a digital camera). They have up to four times the capacity as Nicads and no memory effect. The per-cell voltage is 3.6 volts, and lithium batteries have an exceptional energy density. Lithium batteries require a special charger and can be recharged around 400 times.

Batteries of the Future...

Who knows? The lithium battery is new and expensive now, but may become more common. One of the many other chemistries used in specialty batteries may gain in popularity, or something entirely new could pop up anytime.

"Smart batteries" are also on the drawing board. These have a microchip inside of them to monitor their environment and usage, and manage their recharge cycle.

Cautions

Never try to charge primary batteries. They can give off toxic gasses, explode, or do other unpleasant things.

Don't let your batteries get wet or store them in high-humidity areas.

Don't try to solder or weld directly onto battery terminals, unless the battery was designed to be abused this way.

Don't short-circuit the battery; it could heat up and explode.

When you replace one battery in a pack, replace them all.

Don't mix batteries. In a given battery pack all of the cells should be the same type, brand, age, and grade.

Project 3-1: Motor Power Pack

Except for the 9-volt battery, most batteries are used in a battery pack. Batteries are connected together in packs to increase either their voltage or capacity. See *Figures 3-1A* and *3-1B*.

Qty	Part	Description
1	Battery Pack, AA 4-cell	Series pack, 6-volt output
4	Battery, AA	
1	Header, 2-pin	Vertical 2-pin header, or cut from larger headers
1	LED	Power indicator
1	Resistor, 330 Ohm	Current limit for LED
1	Switch, SPST DIP	One of several, mounted to the prototyping board

Table 3-4. Project 3-1 parts list.

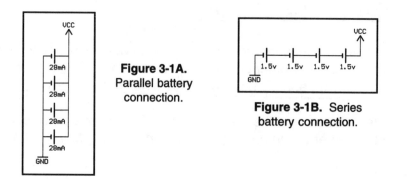

Figure 3-1A.
Parallel battery
connection.

Figure 3-1B. Series
battery connection.

If each battery in a pack is rated at 1.5V and 28mA, then the parallel connected batteries will still have 1.5 volts but four times the current: 4 x 28mA or 112mA. The series connected batteries will only have the 28mA of current available, but at 6 volts. The Tamiya motors use about 4.5 volts, but don't mind a full 6 volts so this project uses a series-connected battery pack. Most battery packs are series connected; if you want more power, you can either user a different battery type or connect multiple packs together in parallel.

The heart of the motor's power pack is the battery holder. The motor's power pack is shown in *Figure 3-2*. You can attach this to the robot's platform with bolts, epoxy, or double-sided foam tape. Only bolts give a reliable attachment. You can insert the (tinned) wires from the battery pack directly into one corner of the prototyping board, or make a more durable plug by carefully soldering the wires to a 2-pin header, which is then inserted into the board.

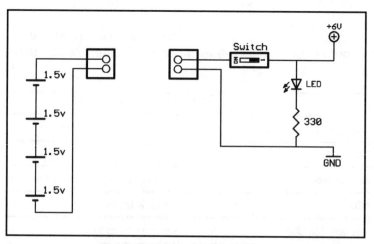

Figure 3-2. Motor's power pack.

The practice of soldering wires onto headers is very handy and I do it to everything. It makes it easier to plug things into the prototyping board, the wires last longer (they don't get scrunched by being plugged in many times), and things look very tidy. It takes a bit of practice before you can do it without short-circuiting everything, or melting the header. You can buy headers in many different pin counts, or you can get the biggest ones and cut them to size (many are designed for just this purpose). For an even classier look, you can get machined socket strips and solder wires into them.

The final housekeeping tip is this: fasten all related wires together using little zip-ties. You can get a hundred of these plastic gizmos for just a few dollars, and once you've snipped the tail off of the zipped zip-tie all of the wires are bundled nice and tidy. Like duct-tape for electronics hobbyists. Great stuff.

Once the battery pack is plugged into the board, the positive lead goes to a switch and the output of that goes through an LED indicator to a positive power bus. The negative lead goes directly to the negative, or ground, power bus. Some prototyping boards don't pass the power busses all the way across the board, but rather split them in the middle. Check the continuity of your board and put jumpers in place if needed.

The switch can be anything, but it can be hard to find a good slide switch that mounts onto a prototyping board. I am using the first switch in a 4-switch DIP.

The LED's resistor value is also not critical. You can calculate it to precisely match the LED's current requirements given the power supply voltage and the LED's voltage drop, or you can stuff in whatever is handy from a 220 ohm to a 1K ohm resistor and get decent results.

Project 3-2: Electronics Power Pack

Most electronics are far more picky about the power supply than the drive motors (though CMOS isn't picky, taking 3 to 15 volts). They tend to require a particular voltage (in our case, 5 volts) and they want the power to be clean and stable. There are two separate issues here: stable power and clean power.

The voltage from a battery decreases over its lifetime, from 1.5 to 0.8 volts (or from 9.0 to 4.8 volts for the rectangular 9-volt batteries). A four-pack of AA cells starts its life at 6.0 volts, but by the time its dead it is only putting out 3.2 volts. This is far from stable, and falls well below the preferred 5-volt limit before the battery is considered "dead." There is a way around this, which is addressed in the next section.

Qty	Part	Description
1	Battery bracket, 9-volt	Little metal bracket to hold the battery
1	Battery clip, 9-volt	Clip with wires that snaps onto the battery's power terminals
1	Battery, 9-volt rectangular	
1	Bolts, 4-40, with nuts	To mount the bracket
1	Capacitor, .1 uf	Power Regulator output filter
1	Capacitor, .22 uf	Power Regulator input filter
1	IC, LM7805 or LM3405-volt Power Regulator	
1	LED	Power indicator
1	Pin Header, 2-pin	For power leads
1	Resistor, 330 Ohm	Current limit for LED
1	Switch, SPST DIP	Shared with Project 3-1

Table 3-5. Project 3-2 parts.

Stable Power: Linear Voltage Regulators

One method to reduce the voltage of a power supply is to run it through a linear voltage regulator, such as the LM7805. For all voltages within the regulator's prescribed limits (a minimum of 7.5 volts to a maximum of 25 volts), the output of a 7805 will be an even 5 volts.

The easiest way to get the necessary voltage to drive the regulator is to use a 9-volt battery. If you have a large enough platform you can wire two 4-battery holders in series, or find an 8-battery holder to provide 12 volts. For some serious power you can use a specialty battery package such as a 12-volt lithium pack for cellular telephones or camcorders, or a battery replacement pack designed for radio control cars. An electronics power pack is shown in *Figure 3-3.*

Whichever battery package you use, attach it to the mobile platform and insert its leads into the prototyping board. Bolt the 9-volt battery clip to the platform.

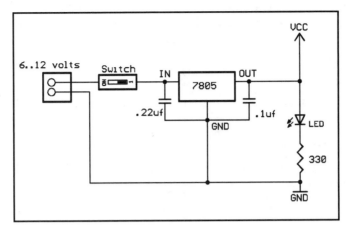

Figure 3-3. Electronics power pack.

Clean Power: Separate Supplies

DC electric motors do terrible things to power supplies. They also do terrible things to controller electronics and even nearby AM radios.

Attach an oscilloscope to the leads of a running DC motor and the trace looks a bit like angora yarn – fuzzy with static. Every time the commutator breaks and makes a connection inside the motor, there is a little arc across the gap. This puts a spike into the power line and into the airwaves. Every time a coil in the motor loses voltage, the magnetic field it was generating collapses and, through induction, creates a reverse voltage back into the driving circuit.

While still watching the oscilloscope, disconnect power from the motor. If the trace looked like angora before, now it's having a seriously bad hair day with static cling. The motor is coasting to a stop, acting like a generator, inductor, and a little spark-gap transmitter all at once. On the little 6-volt toy motors, I've seen 20-volt spikes, both positive and negative. This is bad for unprotected electronics.

This noise feeds back into the power driving the motor, and into any electronics attached to that power. You can reduce the effects of this by decorating the motors like a Christmas tree with capacitors (.1uf capacitor from lead 1 to ground, from ground to lead 2, and from lead 2 back to lead 1; all attached as close to the motor as possible or on the motor itself). Or better, provide separate power supplies for the motors and the electronics.

The robot platform runs off of two separate battery packs, which share only a ground. The motors can dirty up one power source and the electronics run off of the other. Another advantage is that the electronics and the motors can operate from different voltages (for example, 5-volt electronics and 12-volt motors). Motor and electronics power is illustrated in *Figure 3-4*, and power supply parts layout is shown in *Figure 3-5.*

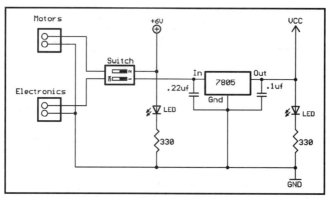

Figure 3-4. Motor and electronics power.

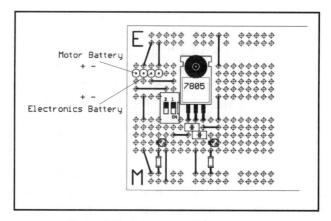

Figure 3-5. Power supply parts layout.

You can combine two SPST switches into a single DPST power switch, but it is handy to have separate switches. This way, you can turn on the electronics and leave the motors off, for testing.

The LM7805 (also sold under the identifier LM340) is a good device for this project, mostly because it is extremely common and it can handle a solid 1 amp. An even better chip is the '2982 or Maxim's MAX8878. The 7805 needs its input to be at least 1 volt higher than its output (it has a 1 volt dropout in my tests, and a 2.5 volt dropout in its specifications). The '2982 has a much smaller dropout and consumes less power when it is unloaded. Check it out as a replacement for the 7805 if you can find it.

Step-Up/Step-Down Regulators

Battery powered applications are different than projects that plug into the wall. You need to be power conscious if you are running off of a battery. They have a limited supply of current to give, so any greedy circuits will run the cells quickly dry.

In addition, battery voltage decays with use—from 6.0 volts to 3.2 volts for four 1.5 volt alkaline cells. Most regulators require an extra volt or two above their rated output before they work correctly, and this shortens the effective life of batteries even more.

There are three solutions to this voltage fade problem: higher-voltage battery sources; 3-volt electronics (of which there is an ample supply), or use a fancier voltage regulator.

Another name for the "fancy regulator" solution is the Step-up/Step-down Switching Power Regulator. These take voltage inputs from 2 volts to 12 and more, and reliably convert them to 5 volts at a respectable current. See Figures *3-6* and *3-7*.

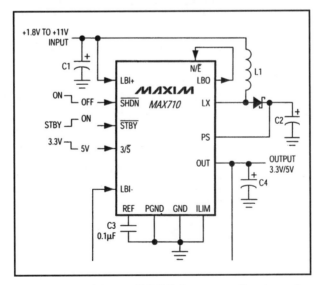

Figure 3-6. Max710 DC-DC Converter. Courtesy of Maxim Integrated Products.

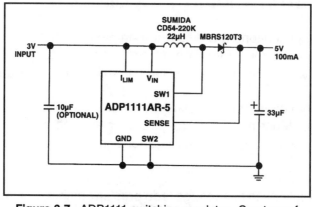

Figure 3-7. ADP1111 switching regulator. Courtesy of Analog Devices.

Three representative chips that perform this function are Maxim's MAX710, Linear Technology's LT1170, and Analog Device's ADP1111.

CHAPTER 4
Electronic Control

There are two worlds within a robot: The mechanical world of unregulated voltages, dirty power, and back-EMF spikes; and the electronic world of clean power and crisp 5-volt signals. You need to bridge these worlds in order to control mechanical systems from digital logic.

The classic component for this is to bridge the relay. A control signal generates a magnetic field in the relay's coil, which physically closes a switch. Relays are useful, yet slow. They have a tendency for their contacts to arc under heavy load – especially inductive loads, which have a certain *momentum* to them and resist being turned off. The collapse of the magnetic field in an inductor, or coil of wire, generates an electrical impulse which can damage a switch or electronic device.

What we use in this book is the MOSFET. MOSFETs are highly efficient silicon switches, available in many sizes like the transistor, and which can operate as a solid-state relay to control the messy mechanical systems.

For low-power projects, the simple logic-level MOSFETs and p-channel devices described in this chapter are fine. There isn't much power going through the system so you don't have to worry about overheating or the other inductive-load problems that become severe with larger motors.

There may come a time, however, when you want to drive a large PMDC motor for a large robot. For such a large project, the value of the MOSFET's "on" resistance $R_{DS(on)}$ makes a huge difference in the heat dissipation of the chip. You also need to worry about junction temperatures within the MOSFET and the coefficients of conduction of the MOSFET package and heatsink. It all gets quite ugly, though many MOSFET manufacturers supply a wide variety of information in the form of data sheets and application notes.

Most large MOSFETs require 10 or more volts potential at the gate to achieve full saturation (though there are some high-power logic-level MOSFETs, such as the IRL3803). The large capacitance at the gate for big MOSFETs (or several medium-sized MOSFETs in parallel) can tax simple drivers. You can not rely on the sink and source current from TTL or CMOS gates to swing the gate quickly enough to avoid overheating.

Before designing a circuit for high-power MOSFETs, research the various specialized MOSFET driver chips. These come in several forms: half-bridge drivers, three-phase drivers, high-side drivers, or just plain voltage multipliers. Designing a high-power MOSFET motor driver from first principles can cause "deep, sleep disturbing problems" (as someone once said), so you may want to buy a commercial solution instead. One quote says that FET stands for "Fire Emitting Transistor."

Project 4-1: MOSFET Switch

Qty	Part	Description
2	Capacitor, .1 uf	Motor filter capacitor
2	Diode, 1N4735	6.2v Zener threshold diode
2	Diode, 1N5817	Switching diode
2	MOSFET, IRLD110 or IRLD014	Logic Level MOSFETs
2	Pin Header, 2-pin	For the motor leads
2	Resistor, 10 Ohm	Gate resistor
2	Resistor, 1K Ohm	Pull-down for the switch
2	Switch, SPST momentary	Small, normally-open pushbutton

Table 4-1. Project 4-1 parts list.

There are two broad families of transistor, bipolar junction transistors (BJT) and field-effect transistors (FET). Both are 3-lead devices, with the flow between two of the leads controlled by a signal at the third lead. In BJT devices, a small current flow at the base moderates a much larger current between the emitter and collector. In FET devices, the presence of an electrical field at the gate moderates the flow between the source and drain. In FETs, there is zero gate current. They act like a capacitor to the driving circuit.

There are two types of BJT devices, NPN and PNP. MOSFETs also come in two flavors, n-channel and p-channel (*Figures 4-1A* and *4-1B*).

N-channel MOSFETs are analogous to NPN transistors (gate instead of base, drain for collector, source for emitter). The drain is more positive than the source and no current flows from the drain to the source unless the gate is raised to a specific voltage above the source.

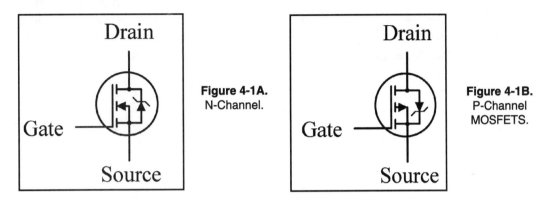

Figure 4-1A.
N-Channel.

Figure 4-1B.
P-Channel
MOSFETS.

P-channel MOSFETs are analogous to PNP transistors. The drain is more negative than the source, and current only flows from source to drain when the gate is brought negative with regards to the source. Though p-channel MOSFETs can be useful in some circuits, they are generally less efficient than n-channel MOSFETs, have a higher on resistance, and are available with less options.

These descriptions apply to enhancement mode MOSFETs. Depletion mode MOSFETs conduct by default and turn off with a gate voltage.

The allowable drain-to-source voltage differential V_{DS} can be in the range of hundreds of volts with drain current I_D ratings available at over 100A. The gate-source control voltage V_{GS} is typically 10 to 20 volts from source (5 volts for devices). At full "logic level" saturation $V_{GS(on)}$ the resistance from drain to source $R_{DS(on)}$ can be as low as 8 milli-Ohms, though typically the lower the I_D rating, the higher the $R_{DS(on)}$ up to a very reasonable 1 to 10 ohms.

R_{DS} starts fairly high at the threshold voltage $V_{GS(th)}$ when the MOSFET first turns on. R_{DS} decreases as V_{GS} approaches the ON voltage, with the MOSFET acting like a voltage-controlled resistor in this "ohmic" region of the device. When using a MOSFET as an on/off switch it is best to charge and discharge the gate as quickly as possible, otherwise the high transitional R_{DS} causes the device to overheat with large loads or frequent on/off switching.

This project uses the MOSFET as a simple on/off switch, one per motor. There aren't any control electronics to hook the MOSFET up to yet, but you can simulate the digital control signal with a small push-button switch. This isn't very exciting, but it will get better soon.

The IRLD110 is a good MOSFET for digitally-driven, low-power switches. It has a gate threshold $V_{GS(th)}$ of 2 volts and saturation at 5 volts. The $R_{DS(on)}$ of .5 ohms also makes it an efficient choice.

Where you place the MOSFET in relation to the load is important. In *Figure 4-2* the MOSFET is on the "low" side. If it was between the motor and the +5v supply it would be on the "high" side. Low-side MOSFETs have simple gate requirements. Since the source is on ground, the gate voltage V_{GS} is calculated from ground.

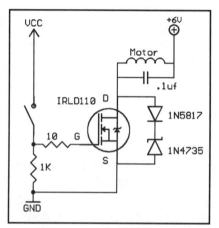

Figure 4-2. MOSFET switch.

If the MOSFET was above the motor, on the high side, things would be different. To drive the gate 5 volts above the source requires a gate voltage that is higher than V_{cc}.

The control logic (the push button and resistor to ground) is a stand-in for now. The resistor to ground pulls the MOSFET's gate low and prevents a short-circuit when the button is pressed. When the button is pushed the gate is pulled high, activating the MOSFET. The resistor's value is not critical. When driving MOSFETs with logic gates, the gate of the MOSFET should never be allowed to float but should have a low-impedance circuit through a totem-poll (push/pull) driver to either V_{cc} or ground. The diode pair across the MOSFET protects it from inductive spikes from the motor. The zener diode sets the voltage threshold for the spikes, and both diodes should have fast switching and recovery times. Most MOSFETs have a built-in protection diode, but this is not always sufficient protection. For these low-power circuits, the values of the components are not critical.

The MOSFET's gate acts like a capacitor, and this is easy to drive. But this capacitor behavior can also be a problem since spikes in the source-drain channel couple back to the gate's driving logic. A strong source/drain spike can actually destroy the driving logic. A resistor in series with the gate helps limit the current of any coupled spikes.

It also slows the switching speed, which reduces the inductive coupling. The gate has a parameter in the data sheets known as gate charge ($C=Q_G$). A rule of thumb for using Q_G is 1 mA of current can switch 1 C of gate charge in 1 micro-second. So, 1 Amp will switch 1 C in 1 nano-second.

Build two of these MOSFET switches shown in *Figure 4-2* and *Figure 4-3*, one for each motor on the platform.

Project 4-2: PWM Speed Control

Qty	Part	Description
2	Capacitor, .01 uf	Decoupling capacitor
2	Capacitor, .1 uf	Timing capacitor
4	Diode, 1N914	Small switching diode
1	IC, 556 Timer	Dual 555 Timer
2	Potentiometer, 100K-Ohm	Pulse ratio control
2	Resistor, 1K-Ohm	Frequency resistors

Table 4-2. Project 4-2 parts list.

Two speeds (on and off) are not sufficient for any kind of decent navigational control. You must have a way to travel at velocities between dead stop and warp nine. The first impulse at speed control is to throw a resistor or potentiometer into the system. Perhaps a different value of resistor for different speeds to moderate the gate signal to the MOSFET, or perhaps in series with the motor at the drain or source. However, extra resistance gives extra heat and a reduction in efficiency. There is a way to change the motor's speed and retain the system's efficiency.

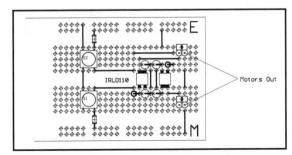

Figure 4-3. MOSFET switch parts layout.

Using pulse width modulation (PWM) we change the percentage of time the motor is turned on instead of changing the current flowing through the motor (*Figure 4-4*). The two extremes of PWM are always-on and always-off, which is what you have with the push-button control from the previous project. In between is where the magic is.

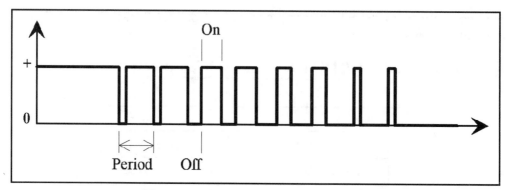

Figure 4-4. Pulse-width modulation signal.

The period is the base frequency of the PWM signal. Too long of a period and control will be rough and the motors will buzz or hum. Too short, and the MOSFET spends more time in transition and can get hot. It's the three-bears problem – not too fast, not too slow, but just right. 10 to 20KHz is a good range for the PWM frequency.

The PWM period is divided between the ON time and the OFF time, and the ratio of ON to OFF determines the speed of the motors.

This circuit uses a 556 dual timer (two 555 timers in one package) to create a dual-speed control (*Figures 4-5* and *4-6*).

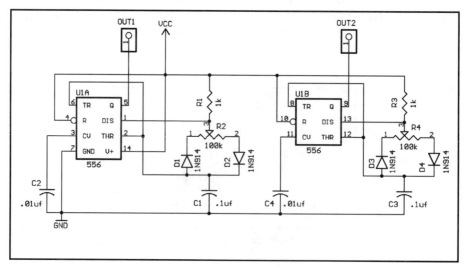

Figure 4-5. Dual PWM controller.

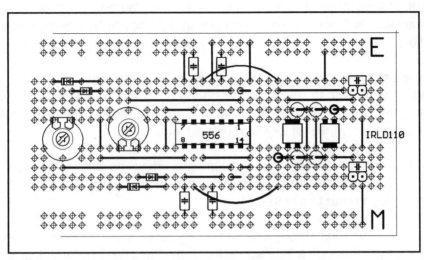

Figure 4-6. Dual PWM controller parts layout.

In a typical 555 timer circuit there are two resistors, R1 between V_{cc} and discharge, and R2 between discharge and threshold/trigger, plus a timing capacitor C to ground. The time it takes to charge the capacitor is:

$$T1 = 0.693 \times (R1 + R2) \times C$$

The discharge time is:

$$T2 = 0.693 \times R2 \times C$$

The frequency of the timer is calculated as:

$$F = 1.44 / ((R1 + R2) \times C)$$

In this timer configuration, the circuit generates duty cycles from 50% to just shy of 100%.

For duty cycles on both sides of 50% you need to break the symmetry of R2, giving the charge and discharge cycles different resistances. This is done by introducing diodes D1 and D2, as shown in the schematic. The capacitor charges through D2 and discharges through D1, and by placing different resistors upstream of the diodes you have full control over the duty cycle. Making those resistors the two branches of a potentiometer, gives manual control over the duty cycle from just above 0% to just below 100%. This circuit will never turn the motor entirely off or entirely on, but it is certainly close enough.

Small potentiometers designed for circuit-board mounting can be found, and they are adjustable with a small screwdriver. For these, and for other electromechanical components, the leads may not fit easily into the prototyping board—they are flat in the wrong direction. In these cases, carefully rotate the leads 90 degrees with a small pair of pliers.

The reset lines of the 556 are tied to V_{cc}, allowing the timers to run as soon as power is applied. If you want an on/off switch on the timers, connect the reset lines to a switch that shifts their voltage from V_{cc} to ground as needed. This reset line can be used as a power-down signal for the motor drivers.

On the parts layout, you can see two jumpers drawn as arcs over the capacitor. These arcs represent air-wires, which go above components on the board because there isn't enough room to run them flat.

Build this circuit and run it through its paces. If you are like me, you'll be down on the floor chasing the robot with a little screwdriver making it run in different circles. If you have an oscilloscope, watch the output of the timer change as you adjust the potentiometer. Though you are still running the robot under manual control, the foundations are laid for automatic control. If you are feeling adventurous, see what happens if you replace the potentiometers with photoresistors or some other variable resistance device.

R/C Servo Control

Although we aren't using R/C servomotors here, it is good to know how to control them since they are extremely useful and fairly inexpensive.

In PWM speed control, the duty cycle of the square wave determines the speed of the motors. In R/C servos, the position of the servo (or, in modified free-running servos, their speed and direction) is determined by a coded pulse as shown in *Figure 4-7.* The period is not critical and is typically 20 mS (milli-seconds, 1/1000 of a second). When the pulse width is about 1.5mS the servo is in neutral (center) position. As the pulse reduces to about 1.0mS the servo moves to its left position. Increasing the pulse to about 2.0mS moves the servo to the right.

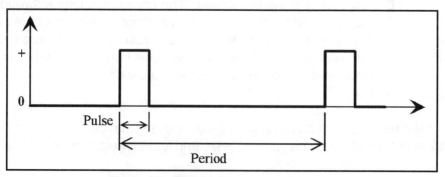

Figure 4-7. RC servomotor control signal.

The servos themselves have three wires – power, ground, and signal. On Futaba servos, red is power, black is ground, and white is signal (though on other brands, this may be different).

If you want to create a servo driver from a 555 timer, R1 is a fixed resistor in series with a potentiometer and R2 is a very large fixed resistor. For the capacitor, choose R1+potentiometer+R2 so that the period is about 20mS, and R1+potentiometer to give a pulse spacing from 1 to 2mS (which is left as an exercise for the reader).

Mechanical H-Bridge

Though PWM control gives different speeds, it does not give different directions. When the robot bumps head-first into a wall, it is important to be able to reverse direction. This means you must be able to reverse the flow of current through the motor, and the circuits so far only switch it on and off.

The circuit that reverses a motor is called an H-Bridge and consists of four switches arranged in an H pattern with the motor in the middle (see *Figure 4-8*). You can find H-Bridge controllers with built-in drivers capable of running a 1-amp load, and one of these is perfect for controlling the little robot platform. You can also find half-bridge controllers, which manage 1/2 of an H-Bridge (one low-side driver and one high-side driver).

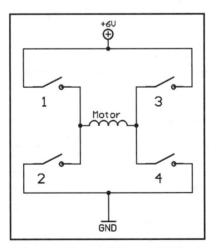

Figure 4-8. Switch-based H-Bridge control.

In a switch-based H-Bridge, pressing buttons 1 and 4 runs the motor one direction. Press 2 and 3 and the motor runs in the reverse direction. Press buttons 2 and 4 or 1 and 3 and the motor doesn't run but the shorted terminals provide a braking action.

Of course, if you press 1 and 2, or 3 and 4, at the same time, you will have a short circuit and something will melt.

Project 4-3: Semi-Mechanical H-Bridge

Adding a 5-volt DPDT relay to the MOSFET switch gives direction control, as shown in *Figure 4-9*. The Dual-Pole, Dual-Throw (DPDT) relay reverses the current through the motor when it is switched. You still have PWM control over the drive MOSFET, and this MOSFET is still switching the low side of the motor. A list of parts for Project 4-3 is shown in *Table 4-3*.

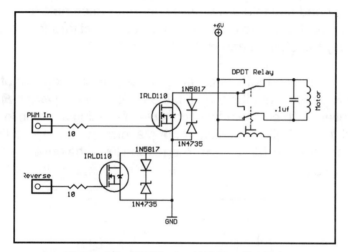

Figure 4-9. DPDT relay H-Bridge Control.

Qty	Part	Description
2	Capacitor, .1 uf	Filter capacitors
4	Diode, 1N4735	6.2v Zener threshold diode
4	Diode, 1N5817	Switching Diode
4	MOSFET, Logic Level	Motor and Relay power switches
2	Relay, DPDT	PC Mount, 5-volt, low-power relay
4	Resistor, 10 Ohm	Current limit on relay

Table 4-3. Project 4-3 parts list.

This type of MOSFET/Mechanical relay combination gives a simple design. You don't have to worry about a high-side MOSFET driver, and you still get PWM speed control with the MOSFET.

There are disadvantages as well, some of which can be managed through careful control. In this circuit you may not want to switch the relay when the motor is going full-speed. Slow the motor down to a full stop before you switch the relay to reduce the back-EMF across the relay. If the motor is going full speed ahead and you reverse the relay, the momentum of the robot will continue to generate electricity – in direct opposition to the new current flow. This is a good way to burn out relays, and possibly anything near them in the circuit. This advice is ignored in later projects. The low current used in these robots creates few problems. Keep the hazards in mind, however, for any larger robots you may build.

Building two of these controllers takes most of the prototyping board, as seen in *Figure 4-10*, so you may want to wait until Project 4-4 before you build this circuit.

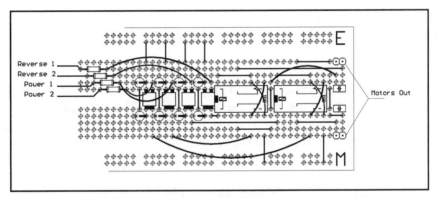

Figure 4-10. Relay H-Bridge control parts layout.

MOSFET H-Bridge

The easiest way to make a MOSFET H-Bridge is to use logic-level p-channel MOSFETs as high-side switches, and the familiar logic-level n-channel MOSFETs as low-side switches. It might look something like *Figure 4-11*. The fun part is finding matched p- and n-channel logic level MOSFETs.

As you review this circuit, bear in mind that n-channel MOSFETs have their source at ground, their drain on the load, and conduct when the gate voltage is high. P-channel MOSFETs have their source on the load, their drain at V_{cc}, and conduct when the gate is low. We are still using logic-level MOSFETs, so V_{cc} and $V_{GS(on)}$ voltages are both 5 volts.

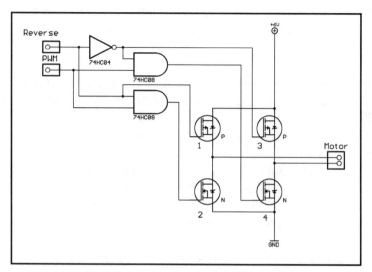

Figure 4-11. P-channel H-Bridge controller.

The logic gates combine the Reverse and PWM control signals, and drive the MOSFETs accordingly. These gates also prevent any illegal configurations of the MOSFETs.

With this circuit it is still the best policy to stop the motors (PWM off) before switching the reverse line, otherwise there will be a short time when all four MOSFETs are conducting at once, and the resulting near-short can cause problems. Specialized H-Bridge controller chips automatically delay between turning off one half of the bridge and turning on the other half, known as dead time.

Project 4-4: Permanent Motor Drivers

In order to free-up the prototyping board for other uses, you need to build the motor controller circuit on a separate board. There are different ways to do this – you could stack a second prototyping board above the first, leaving the power supplies and motor drivers on the lower board. You could purchase a circuit board version of the prototyping board, solder the components to it, and stack that. You could even place two prototyping boards side-by-side on the robot to give more room, if the robot is wide enough. A list of parts for Project 4-4 is shown in *Table 4-4*.

	Part	Description
1	PC Board, General Purpose	Almost 2" by 3", with at least 25 holes along the edge
4	Pin Header, 2-pin	Signal and Motor connectors
2	Pin Header, 3-pin	Power connectors

Table 4-4. Project 4-4 parts list.

This project uses a small circuit board, shown in *Figure 4-12*, that is exactly the right size to hold two of the H-Bridge controllers described in Project 4-3.

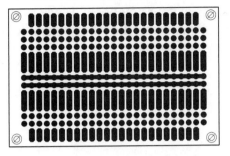

Figure 4-12. Small project circuit board.

Using this board plus all of the components described in Project 4-3, solder together the motor drivers shown in *Figure 4-13*. Take all 12 wires leading from the board and solder them to pin-headers. The power lines go to the outside pins of 3-pin headers, clipping the inside pin off. This can then be plugged (with some fiddling) into the power busses from the prototyping board. The control lines go to 2-pin headers which plug into the edges of the prototyping board.

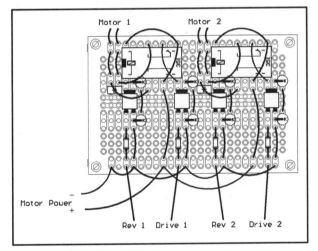

Figure 4-13. Relay H-Bridge, permanent parts layout.

Mount the driver board on the bottom of the robot platform. *Figure 2-11* shows a good location for this board between the motors, and over a pair of feeler switches attached in a later project.

All of the wires should be color-coded or, failing that, you can use white correction fluid to mark the bodies of the headers. There is nothing more confusing than a mass of colorless, undifferentiated wires a few weeks down the road. Be sure to document the marking scheme you use in your notebook! Black is usually ground, Red power, and White makes a good signal wire. But what colors do you use for Reverse 1 or Drive 2? Good notes are your friend.

CHAPTER 5
Adding Sense

The previous chapter left you with several circuits for controlling the robot. This chapter assumes that the robot platform has the PWM speed control from Project 4-2 driving the H-Bridge built in Project 4-4.

Now that there is full electronic control over the robot's motors, it needs to be hooked up to some sort of brain. But how do the brains decide what actions to take, in a complete absence of sensory input? It is vital that the robot be able to interact with the world – not only to affect its surroundings, but to sense them as well.

This chapter introduces a variety of senses for the robot and illustrates how it can use them to interact with its world in a rudimentary way.

Project 5-1: Time

Qty	Part	Description
1	Capacitor, .01 uf	Decoupling capacitor
1	Capacitor, .47 uf	Timing capacitor
1	IC, 555 Timer	Timer
1	LED	Pulse indicator
2	Resistor, 1M Ohm	Timing resistors
1	Resistor, 330 Ohm	LED current limiter

Table 5-1. Project 5-1 parts list.

Why would a robot need a sense of time? I don't mean time such as the hour of the day (which can also be added, with the addition of a peripheral clock chip such as the DS1629), but a sense of rhythm, a pulse, a mental metronome.

Things take time. Events impinge on the robot in a blur and the robot is trying to sort the chaos of the world into an orderly progression of events. You bump a wall, you need to back up... but how far? How much time do you spend backing up? A robot with a rapid sense of time may back up a very short distance... and go forward and run into the wall again, and back, and then forward... frantically tapping its way to a clear path. A slower robot may do a leisurely retreat before going ahead again, describing a much different path and reflecting a much different personality.

The sense of time provides the basic divisions which the robot reacts in. Change this time base, and you change not only the robot's behavior but its personality. In most systems this sense of time is implied by the various circuits and programming. It suits our development plan to make it explicit. See *Figures 5-1* and *5-2*.

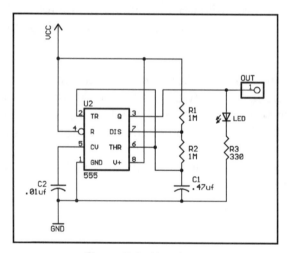

Figure 5-1. Heartbeat.

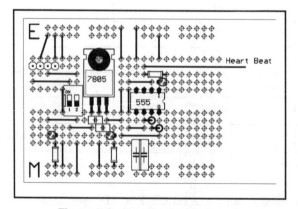

Figure 5-2. Heartbeat parts layout.

This is the same type of timer circuit used in the PWM and R/C pulse generators. If you want to conserve space, you can combine circuits and use the 556 or 558 timers. The 558 has four 555 timers on one chip, and the 556 has two timers.

Though the 555 is technically a TTL chip, you can find CMOS versions of it. These are better because they have lower power consumption, add less noise on the bus, and provide better output voltages.

The values of R1, R2, and C1 are up for grabs. Change them as you wish to adjust the heartbeat of the robot. The duty cycle isn't important, but remember that R1 should be no less than 1K, and R1+R2 no more than 3.3M.

Project 5-2: Touch

Qty	Part	Description
1	Capacitor, 20uf	
1	Capacitor, 33pf	
1	IC, 74HC14	Schmitt gate
1	IC, LM324	Operational Amplifier
1	Microphone	Electret Microphone Cartridge
1	Potentiometer, 100K Ohm	Threshold Adjustment
1	Resistor, 100K Ohm	
1	Resistor, 1K Ohm	
1	Resistor, 1M Ohm	
1	Resistor, 2.2K Ohm	

Table 5-2. Project 5-2 parts list.

Touch, for us, is a most intimate sense. Every speck of our body is sensitive to contact, and these signals help us to walk, pick things up, and even feel our way around in a darkened room. With a simple sense of touch the robot will also be able to feel its way around what, to it, is the darkest of rooms.

The robot will have a very crude sense of touch... a pair of simple on/off switches at the front of the platform. Find a switch with a long lever like the one in *Figure 5-3*, which can be mounted on the robot in the position shown in *Figure 2-11*. Note that the lever is facing into the centerline of the robot... and in that position there is no way that it could reasonably be activated. Cut the switch's lever back so it pokes out only so far as to allow you to attach the feeler to it. Also put insulating tape over the switch before mounting the driver board over it, to keep it from shorting out the drivers.

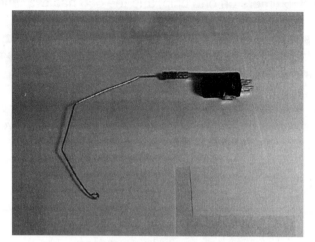

Figure 5-3. Switch for feelers.

Next, shape and attach a long wire (like a bug's antennae or feeler) to the lever of the switch. The wire should be long and curve out and to the sides of the robot like in *Figure 5-4.* The prototype used 0.040" (16 gauge) stainless steel wire, which may be hard to find (I bought it in 25 lb. spools for another project). Whatever wire you use must be tough and resilient. It is going to be smashed into obstacles at whatever speed your robot drives at. The momentum and slow reaction time of the motors will guarantee that the feeler will receive more than a gentle touch.

Figure 5-4. Position of feelers.

The easiest way to attach the wire to the switch is with the help of crimp-on quick-disconnect connectors. Find a "female" connector that fits snugly over the switch's lever, and crimp the feeler wire into it. It helps to bend the end of the wire over double so it makes a snug fit into the connector. For extra strength, solder the connector around the wire after it is crimped, but be careful not to get solder into the "quick" part of the disconnect. Once this is done, push the connector onto the switch. An advantage of this scheme is that when the feelers are bent and mangled beyond use, they can be easily replaced.

Feeler switches, by themselves, do the robot little good. *Figure 5-5* shows a circuit that uses their signals.

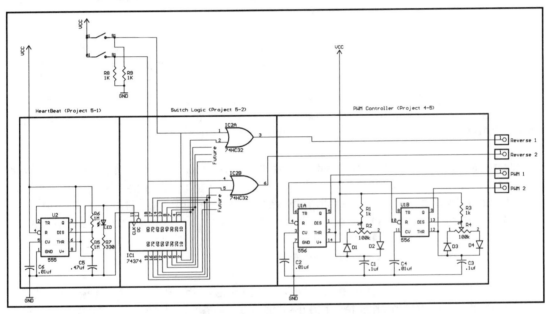

Figure 5-5. Control logic parts layout.

The schematic in *Figure 5-5* and board layout in *Figure 5-6* show the switch control logic in context. The left block is the heartbeat developed in project 5-1, and the right block is the PWM controller developed in project 4-5. Not shown are the power supplies and the motor driver board, which are in their usual locations. The plugs from the driver board, switches, and motors are plugged into the prototyping board as shown in the parts layout.

The switch logic receives the heartbeat signal from that circuit and information from the switches. It sends out control signals directly to the reverse line of the motor controller board. That is all this brain does, however you can also control the drive signal from the PWM by using the reset lines of the 556 timer. A high value on the reset allows the PWM to generate a signal; a low signal turns off the timer and, by extension, disables the motor drive.

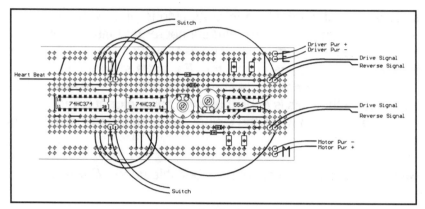

Figure 5-6. Control logic parts layout.

Now to describe the reflex system that gives the robot its first behavior. As the robot is slowly driving along the floor (or, in my case, madly careening across the kitchen) it will eventually bump into something. This bump closes the contacts in one or both switches, creating a signal that the robot can act on. If the robot has rammed headfirst into a cabinet, the switch will go on and stay on until the condition is relieved. If the robot brushes along an obstacle, the switch may create only a very brief signal.

You can hook the output of the switch directly to the reverse line on the motor driver so that when the left switch hits an obstacle the left motor reverses. This ignores, for this project, the hazards of reversing the motors while running. Though not the best way to do things, it works well enough for now. This quickly releases the contact on the switch – too quickly, in fact, to make much of a difference and the robot will immediately ram into that obstacle again.

This circuit feeds the output from the switch into one of the 8 latches in the 75HC374. The output of the first latch is fed into the input of the next, and so on down the line — four latches in sequence after each switch. When the switch is activated during a clock transition or is held closed long enough for the clock to latch it, the signal is passed one latch at a time down the line until it falls off the end. This gives five opportunities to do something with a switch input. Once at the switch itself plus four more signals from the outputs of the four latches.

For a simple test, the switch and the output of the first latch are fed to the reverse line of the motor on the same side as the switch. The other three latches are ignored. This causes the robot to briefly reverse the motor immediately after the switch is closed and, if this tiny correction does not clear the switch before the next heartbeat, the signal is stretched out by an additional heartbeat, forcing a longer reverse motion (even after the switch is released).

With five signals to work with, there are many options for hard-wired behaviors. You can combine the signals through logic gates or use 4-of-16 decoders to drive the

PWM reset line and/or the motor reverse. With enough prototyping board space and some imagination, there are literally thousands of patterns the robot could perform with just two switches and a sequencer.

A technical note about the 74HC374: this is an edge-triggered latch, which is an important distinction and different from a level-triggered latch like the 74HC373. In an edge-triggered device, the data is latched only when the clock pulse changes from (in this case) low to high. In a level-triggered device, the latch outputs are allowed to change during the entire time the clock pulse is high (or low). This allows the data to run through the sequencer in an instant instead of at the stately pace dictated by the heartbeat.

Other Feelings

Simple on/off feelers are useful yet crude. Improved forms of touch can give values between all-off and all-on. Potentiometers, for example, give different resistances depending on how far their shaft is rotated. Linear potentiometers do the same thing with a slide bar. A crude home-built force feedback device can be made from conductive foam sandwiched between two metal sheets. There is also a large (and expensive) variety of pressure, torsion, tension, and stress sensors available on the market.

Project 5-3: Light

Qty	Part	Description
1	Diode, Photo	Photodiode
1	IC, Op-Amp	Operational Amplifier
1	IC, Schmitt Trigger	Schmitt Trigger gate or inverter
1	Potentiometer, 100K	Zero-set potentiometer
2	Resistor, 100 K Ohm Photo	CdS Photoresistor
2	Resistor, 1K Ohm	
1	Resistor, 4.7K Ohm	
1	Transistor, Photo	Phototransistor

Table 5-3. Project 5-3 parts list.

After the feelers on the robot are smashed and twisted beyond recognition, it is reasonable to start sensing the environment from a safe distance. Most of us humans are able to navigate around our environment using light reflected from the objects around us and use the sense of touch as more of an emergency backup system.

The ability to sense light is found not only in most animals, but also in plants and almost every other living organism. And soon a primitive version of sight can be found in this robot.

Light causes ionization in silicon. This is used directly in photovoltaic (solar) cells that output a voltage in response to light (like a "light battery"), or indirectly in detectors like the photodiode or the phototransistor. Light changes the resistance of the material in a photoresistor, from high resistance (100K to 1M) in the dark to very low resistance in bright light. Light sensors are shown in *Figure 5-7.*

Photodiode   Phototransistor   Photoresistor

Figure 5-7. Light sensors.

Phototransistors are fairly sensitive and can provide a nearly full-range signal level from 0 to 5 volts when used as shown in *Figure 5-8.* This signal can be fed to an op-amp (to amplify the signal), comparator (to compare it to a threshold level), Schmitt trigger (to digitize it), or to an analog-to-digital converter (to quantify it). Many phototransistors are especially sensitive to infrared light, though they will respond to other frequencies as well.

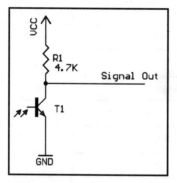

Figure 5-8. Phototransistor circuit.

A similar circuit is possible with the Photoresistor, as shown in *Figure 5-9.* This circuit is a voltage divider whose output is:

$$V_{OUT} = V_{IN} \times (R2 / (R1 + R2))$$

When R1 is in full light it's value approaches zero, giving the simpler equation:

$$V_{OUT} = V_{IN} \times R2 / R2$$
$$= V_{IN} \times 1$$

When R1 is in full darkness, its value is at maximum resistance. This resistance should be at least 10 times the value of R2, and R2 needs to be large enough to limit the current drain when R1 goes to zero. Using the values of 100K for R1 and 1K for R2 gives:

$$V_{OUT} = V_{IN} \times (R2 / (R1 + R2))$$
$$= V_{IN} \times 1,000 / 101,000$$
$$= V_{IN} \times .01$$

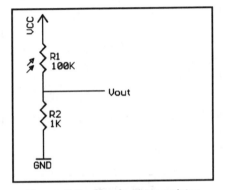

Figure 5-9. Simple photoresistor voltage divider.

Many other sensors change their resistance in response to a stimulus, so the principles and circuits developed for the photoresistor apply to them as well.

Of course, the robot may not be operating in such a dynamic environment that the sensor swings all the way from ground to +5 in the course of its wandering. A sensitive circuit that detects small changes in resistance, while rejecting any ambient or base level resistance, is the Wheatstone bridge shown in *Figure 5-10*. Its output is based on the ratio of all four of its resistors:

$$V_{OUT} = V_{IN} \times ((R2 / (R1 + R2) - (R4 / (R3 + R4)))$$

The Wheatstone bridge circuit is essentially two voltage dividers. The R1/R2 half contains the sensor and the R3/R4 half allows you to zero-set the circuit. In the photoresistor example, place the robot in neutral light (neither light nor dark) and adjust the potentiometer R3 so that the voltage across the outputs is zero. As the light level changes across R1, the outputs will read plus or minus. The Wheatstone bridge gives a differential signal. The value of "Signal +" is read with respect to the value of "Signal −,"and it can swing above and below "Signal −."

The bridge can fix any sensor offset, forcing the output to start from zero. This does not help if you have a small sensor range, where the resistance of the sensor changes a small amount, giving a small signal. To solve this, the signals can be fed into an amplifier. This is shown in simplified form in *Figure 5-10*. Sensor amplification is explored in a later section.

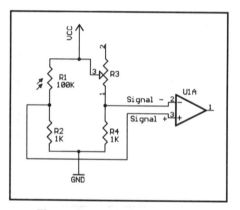

Figure 5-10. Photoresistor in a
Wheatstone Bridge.

For the robot, common off-the-shelf CdS photoresistors can be placed directly into the PWM circuit to adjust the motor speed in response to light, as shown in *Figure 5-11*. In the naïve implementation shown here, the entire potentiometer and diode complex is removed and the photoresistor is put in its place. Soldering long leads to the CdS cells lets you place them in front of the robot. Make a small roll of tape around the sensor's body, to limit its vision to whatever is directly in front of it. One way of doing this is to take a short length of tape (I prefer silver duct tape) and wrap it around a pencil or other round object with the sticky side out. This is pushed over the photoresistor, and the assembly is then stuck to the robot's platform.

Attach the left-hand photoresistor to the PWM timer for the left-hand motor and the right-hand photoresistor to the right-hand motor, and see what it does. Test this arrangement in a slightly darkened room with a bright flashlight to guide the robot for best results. Reverse the signals (left sensor to right motor, and vice versa) and test the opposite behavior.

There are other variations you can try. Leave the diodes in place and wire the photoresistor into only one branch of the potentiometer circuit, replacing the other branch with a fixed resistor. Then try the sensor in the other branch. Use four photoresistors in different configurations on the robot, one for each diode. Point some of these up instead of forward, or point some backwards.

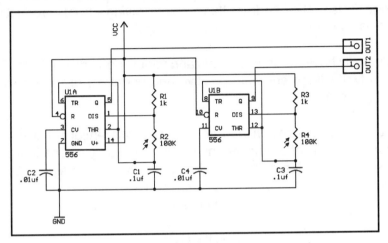

Figure 5-11. Photoresistor in the PWM circuit.

Instead of sensing ambient and naturally reflected light with passive sensors, you can sense obstacles by adding a head-light next to the sensor. Light reflected from objects is detected by the sensor. Of course, you need to shield the sensor from the light next to it or it will be flooded regardless of the presence of obstacles. If you are using infrared sensitive phototransistors, you get the best results by lighting with an IR LED. This can be hard to calibrate since the IR light is invisible to the naked eye.

The analog signal from the light sensor can be digitized by passing it through a Schmitt trigger as shown in *Figure 5-12*, and the digital result used like contact switches. The value of the fixed resistor in the voltage divider can be changed to adjust the trigger point.

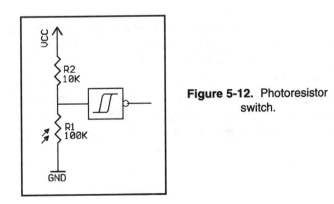

Figure 5-12. Photoresistor switch.

Signal Amplification

The light sensors in the previous section were kind enough to provide a large resistance swing across their normal working range. Other sensors, such as the microphone in the next section, are not as easy to work with and give small signals in the milli-Volt range. This section address this problem, giving a brief introduction to signal amplification using operational amplifiers (op-amps).

Op-amps generally have two inputs – plus and minus, add and subtract. They can be powered from two voltage sources (positive and negative) or from a single source (positive and ground).

The classic op-amp circuit shown in *Figure 5-13* is a differential amplifier and is useful for amplifying the results of a Wheatstone bridge, for example. The input resistances R1 and R2 should be at least 10 times the impedance of the signal, to prevent the amplifier from affecting the source circuit. You can also simply select values that are convenient, if not as effective. The amplification of this circuit is given by:

$$R1 = R2$$
$$R3 = R4$$
$$V_{OUT} = (V2 - V1) \times (R3 / R1)$$

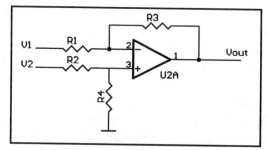

Figure 5-13. Differential amplifier.

Without any feedback resistors, the op-amp functions like a digital switch. The amplification is so extreme that it snaps to full on or full off. A circuit like that shown in *Figure 5-14* compares the levels of two photoresistors, gives a high output if the "+" sensor has more light then the "−" sensor, and vice versa. This type of comparator circuit has the added benefit of eliminating the effects of ambient light. It is the difference in light that controls the output, not the absolute levels. If you want this comparison behavior, a comparator like the LM339 is a more appropriate component.

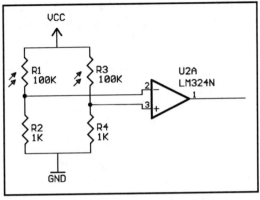

Figure 5-14. Which photoresistor has more light?

The simplest op-amp circuit uses only the positive input of a single amplifier as shown in *Figure 5-15*. Amplifier schematics don't normally show the power connections, but in *Figure 5-15* I wanted to point out the decoupling of the power supply with a pair of capacitors. C1 should be 0.1uf and C2 47uf. The amplifier used for these examples is the LM324 quad, low-power op-amp, though almost any op-amp would work (the LM741 is quite popular).

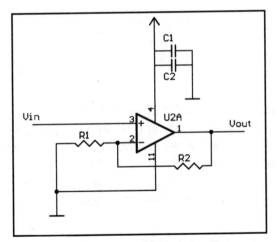

Figure 5-15. Single input amplifier.

The input signal's amplification is based on the values of R1 and R2:

$$V_{OUT} = V_{IN} \times (\, 1 + (R2\,/\,R1)\,)$$

To convert a 50 mV signal to a 5 volt signal requires a 100-fold amplification, which is reasonable for a single-stage amplifier.

For the extreme amplification needed by some other projects, the three op-amp instrumentation amplifier shown in *Figure 5-16* comes to the rescue. This amplifier uses three of the four op-amps in the LM324.

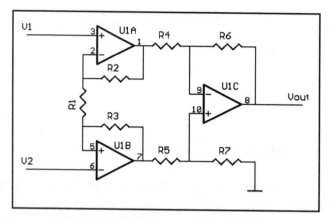

Figure 5-16. Instrumentation amplifier.

The signal is amplified twice, first at the input amplifiers and again at the output amplifier. The total amplification is given by:

$$R2 = R3$$
$$R4 = R5$$
$$R6 = R7$$
$$V_{OUT} = (V2 - V1) \times (\,(\,1 + (2{*}R2\,/\,R1)\,)\,) \times (R6\,/\,R4\,)\,)$$

The inputs V1 and V2 feed directly to the input amplifiers which provide a high impedance input. Gain can be adjusted by splitting R1 into two parts, a potentiometer R1A, plus a fixed resistor R1B to keep the resistance from going to zero. Due to imperfections in the resistors and amplifiers, the plus and minus inputs will not be perfectly matched causing a DC offset in the output. This is fixed by splitting R7 into two components, R2A and R2B, to provide common-mode rejection. This is all shown in *Figure 5-17.*

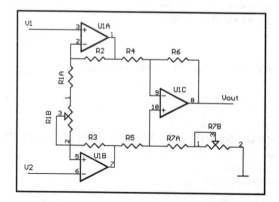

Figure 5-17. Gain and common mode rejection control.

Many design hassles can be avoided by using specialized amplifiers designed for the task at hand (such as the LM386 audio amplifier offering gain from 20 to 200 with few external parts), assuming one can be found.

Project 5-4: Sound

Qty	Part	Description
1	Capacitor, 20uf	
1	Capacitor, 33pf	
1	IC, 74HC14	Schmitt gate
1	IC, LM324	Operational Amplifier
1	Microphone	Electret Microphone Cartridge
1	Potentiometer, 100K Ohm	Threshold Adjustment
1	Resistor, 100K Ohm	
1	Resistor, 1K Ohm	
1	Resistor, 1M Ohm	
1	Resistor, 2.2K Ohm	

Table 5-4. Project 5-4 parts list.

A sense of touch, a sense of sight... what's next? We don't have the technology to implement smell or taste on the market yet, though there are some exciting devices in research labs. This leaves sound! The ability to detect and act on the subtle vibrations of the air around us.

Using an omnidirectional electret condenser microphone cartridge is simple, as seen in *Figure 5-18*, but the signal it produces is small—only 1/10 of a volt peak-to-peak for a loud whistle right at the microphone. For detecting sound, there is no escaping the need for a signal amplifier such as the one shown in *Figure 5-19* to crank the tunes up to 5-volts peak-to-peak.

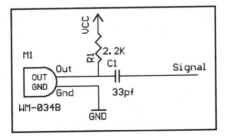

Figure 5-18. Electret microphone cartridge circuit.

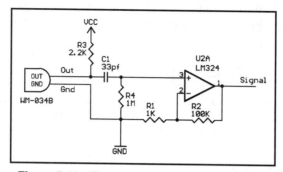

Figure 5-19. Electret microphone and amplifier.

This is an unsophisticated circuit, certainly no good for audiophile applications, but it gives a signal that can interface with 5-volt circuitry. When silent the output sits at a DC offset of about 2 volts, and it is perturbed up and down by sound. This signal can be sampled by a fast A/D converter to give a history of the signal. What this simple robot needs is a way to detect the presence of loud sound as opposed to its absence. The actual waveform isn't useful.

With a little extra processing, the amplified signal can be massaged into a digital pulse as shown in *Figure 5-20*. This circuit is a bit of a hack, and none of the component values are critical. The second amplifier is operating as a comparator. It snaps to ground when the input is below the voltage set by the potentiometer, and switches to V_{cc} when the input is above. C2 then integrates this square wave into a smooth pulse, which is then cleaned up by the Schmitt trigger.

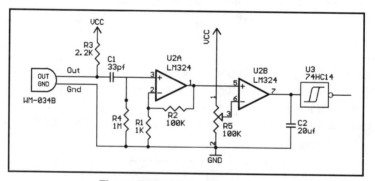

Figure 5-20. Microphone switch.

CHAPTER 6
It's Alive! Simple Robot Behaviors

By now you have a set of robot tools and circuits covering the entire worktable. These are the pieces of a puzzle. These projects have stitched together some muscle, a heart, eyes and fingers. When electricity is applied, they twitch and react; but so does a flayed frog's leg, as demonstrated by early experiments in galvanic response.

Somehow the forced reaction of frog muscle to electric current is different from the actions of the live frog. What makes the difference? Avoiding the difficult issues of spirit and soul, one observable difference is the appearance of intention. The living frog acts and reacts in its environment in a way that benefits the frog. It demonstrates intentional behavior.

A frog has simple needs... stay damp, keep cool, avoid predators, eat bugs, find attractive frogs when the spring breeze plays romantic tunes through the vibrant swamp reeds, etc. Everything the frog does supports the frog's needs. What needs do small, wheeled robots have?

A synthetic organism has synthetic needs, and this is addressed in the classic 1984 work by Valentino Braitenberg in his book *Vehicles: Experiments in Synthetic Psychology*. In that book, he describes 14 increasingly complex mobile robots (vehicles), their behaviors, and the biological rational behind them. MIT uses these behaviors in their Autonomous Mobile Robotics course and with their LEGO™ Brick projects.

Project 6-1: Braitenberg Behaviors

Qty	Part	Description
1	IC, LM324	Operational Amplifier
2	Resistor, 100K CdS	Photoresistor
2	Potentiometer, 100K Ohm	Threshold Adjustment
2	Resistor, 1K Ohm	

Table 6-1. Project 6-1 parts list.

Vehicle 1

This simplest class of creatures is described as having only one motor and one sensor. Steering is accomplished through the whims of friction and the perturbations of the floor surface. The sensor merely adjusts the speed of the motor in response to some stimulus.

Timid

A timid robot only moves when the light falling on its upward-pointing sensor is above a particular threshold, and it stops again when it is in the shade. Build the circuit shown in *Figure 6-1* and *Figure 6-2*. Set the robot in the shade, turn it on, and adjust the threshold potentiometer until the motors just turn off. Whenever the light on the sensor goes above that level, the robot will run until it finds shade again. Chase the robot around with a flashlight.

You can make this circuit partially self-adjusting by replacing the threshold potentiometer with a forward-facing light sensor. Then the only time the robot stops is when the forward sensor detects more light than the upward sensor.

This circuit and the later ones in this chapter, only drive the on/off line on the motor controller. Be sure to ground the forward/reverse lines so that the control MOSFETs are not floating. Floating MOSFETs can trigger unpredictably.

Indecisive

Instead of start/stop control, the sensor can control forward/reverse, driving forward into the light but then backing up into the dark again. Ultimately sitting at the shadow boundary, unsure of what to do.

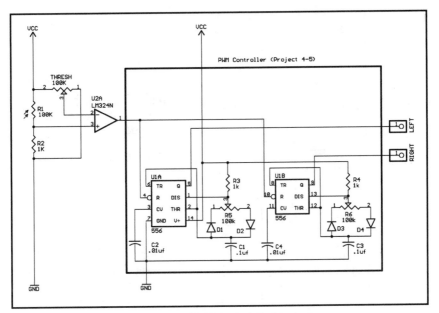

Figure 6-1. Vehicle 1: A timid robot.

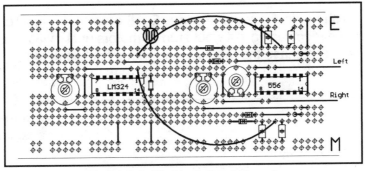

Figure 6-2. Timid robot parts layout.

Vehicle 2

Fear

This second class of creature has two sensors that stimulate two motors. These can be connected in two different ways as shown in *Figures 6-3A* and *6-3B*. With connections to the same side, stimulus on the left sensor activates the left motor and stimulus on the right sensor activates the right motor. The robot will shy away from light, appear to be "afraid" of lights.

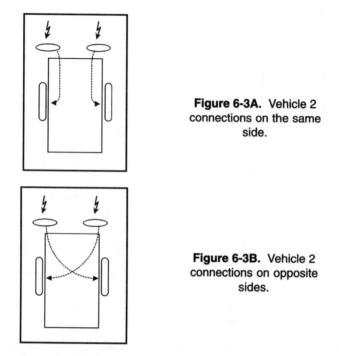

Figure 6-3A. Vehicle 2 connections on the same side.

Figure 6-3B. Vehicle 2 connections on opposite sides.

Build the system shown in *Figures 6-4* and *6-5*. If there was a bright light on the floor and you placed the robot facing it, it would run its motors until it was facing away from the light. You can also chase the robot around with a flashlight, tormenting it with its fears.

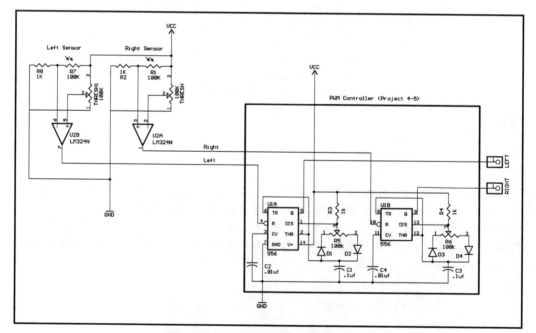

Figure 6-4. Vehicle 2: Afraid of Light?

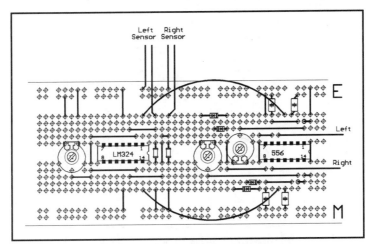

Figure 6-5. Frightened robot parts layout.

Aggression

With the sensor/motor connections reversed, the robot will turn toward the light and accelerate directly into it, possibly even destroying it. Once the eyes are in darkness again, both motors stop moving. They only come to life again when aggravated by lights.

Vehicle 3

Love and Explore

Where vehicle 2 stimulates the motors, vehicle 3 inhibits them as shown in *Figure 6-6*. This vehicle runs both motors in the absence of light, questing for the illumination they crave. When the robot shown in *Figure 6-7* is in the dark, it drives ahead. When both sensors are in the light, it stops. Light and darkness are still defined by the setting of the threshold potentiometer.

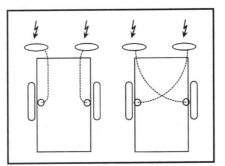

Figure 6-6. Vehicle 3 connections.

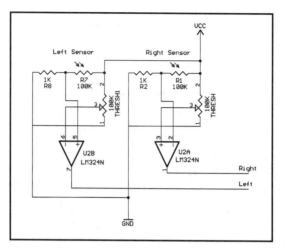

Figure 6-7. Vehicle 3; quest for light.

The same-side inhibition version steers the robot until it is pointed at the light. The crossed connection version will also face the light, but perhaps at an angle, waiting as it were for another light, satisfied for now but poised and ready for another light to approach.

The simple on/off control of these circuits perform their actions somewhat crudely. A variable speed control gives more elegant behavior. Instead of driving the reset line on the timers, a light sensor can be connected to the control voltage (CV) of the timer (refer to the data sheet for the 556 timer to follow along). Both the threshold (pins 2 and 12 on the 556) and trigger (pins 6 and 8) are run through a comparator inside the timer that can be adjusted through the control-voltage line (pins 3 and 11). Inside the 556 is a 3-resistor voltage divider, with the trigger compared to $1/3\,V_{cc}$ and the threshold at $2/3\,V_{cc}$. The control voltage pin ties into this network at the 2/3 point. If you force the CV to be above $2/3\,V_{cc}$, the timer has more "high" in the output; if you force CV near ground, the output goes entirely low. Using the CV pin, a low value from a sensor can halt the PWM control to the motors, while an exceptionally high value will make them drive faster than their ordinary set-point.

Enough for Now

These first few vehicles are simple enough to build using only a handful of components. However, every time you want to try something new, you need to move wires, place chips, adjust, tweak, and tune. All of this takes time and precious prototyping board space. But there is a better way.

There are more vehicle behaviors and other exciting experiments waiting to be performed once you have the big guns put into place.

CHAPTER 7
Microcontroller

Before you start this section, jump ahead to Project 7-1 and order the parts you need. It takes a few days to get them, so it's best to get an early start.

Microcontrollers (MCU) are like the microprocessor (central processing unit, or CPU) inside large home computers. MCUs are slow and can address less memory than CPUs, but they are designed for real-world control problems and are both inexpensive and easy to use. One of the biggest differences between CPUs and MCUs is the number of external components needed to operate them. MCUs can often run with zero external parts, and typically only need an external crystal or oscillator.

Some of the standard CPUs can be found in MCU packages, with the same speed and instruction set as their desktop form, but repackaged for low external component count plus some additional I/O control features.

There are four basic aspects of a microcontroller: speed, size, memory, and options.

Speed is designated in clock cycles, and it is usually measured in millions of cycles per second (Mega-Hertz, MHz). How different MCUs use these cycles varies, and this makes a difference in the useable speed of the processor. Complex Instruction Set Computers (CISC) use anywhere from 1 to dozens of clock cycles per instruction executed, and it can be difficult to determine exactly how "fast" they are. Reduced Instruction Set Computers (RISC) have fewer and simpler instructions, and execute each instruction in one or two cycles. Of course, it may take more RISC instructions to perform the action of a single CISC instruction. Both types of architectures get the job done, using different philosophies.

Size specifies the number of bits of information the MCU can process in one step— the size of its natural cluster of information. MCUs come in 4, 8, 16, and 32 bits, with

8-bit MCUs still being the most common. Most home computers are 32-bit machines. The 64 in "Nintendo-64" refers to the 64-bit data pipeline its graphics processor uses. Memory in desktop computers is measured in millions of bytes (MB), but these small microcontrollers count most of their ROM in thousands of bytes (KB) and RAM in single bytes. Microcontrollers use their memory a bit differently than a desktop machine. Microcomputers organize their memory in a Von Neuman architecture, where programs and data are all stuffed into the same memory system. Many MCUs use the Harvard architecture, where the program is kept in one section of memory (usually the ROM, Flash, or other non-volatile system) while data and temporary variables are managed in a separate second section of memory (usually the internal or external SRAM). This is not only a convenient functional split (permanent data in ROM, temporary data in RAM), but it can allow the processor to access the separate memories more efficiently.

Survey of Microcontrollers

There is a bewildering array of microcontrollers on the market, and finding an MCU for a project can be an intimidating process. It seems that every company makes one or more MCUs (sometimes hundreds), both custom and in one of the "standard" families. A few of these families, and the options they come with, are surveyed here. There are many other MCU types and quite a few unmentioned features. The ones listed are those that are the most common, as well as those that you are most likely to use in other projects.

Memory

Desktop computers keep programs on magnetic hard-drives, floppies, and optical CD-ROMS. It executes those programs out of Dynamic Random-Access Memory (DRAM). Microcontrollers hold their programs in different types of memory. Instead of the volatile DRAM (which loses the program when power is lost, and which must be refreshed regularly) MCUs keep their code in non-volatile memory, since they usually don't have the benefit of magnetic off-line storage.

ROM

Read-Only Memory (ROM) lets you read information out, but you cannot write any back in. Neither can you erase the data in the ROM. Plain-vanilla read-only memories are programmed at the manufacturer from a ROM mask provided by the customer. If you aren't making thousands of these, the manufacturer isn't interested. This system of programming is fine for the MCUs in cars, microwave ovens, and other consumer items, but it isn't very useful for the hobbyist.

Large companies use special In-Circuit Emulators (ICE) to develop the software for these ROMs. ICE tools plug into the socket where the MCU would go and simulate the MCU, communicating with a host computer to get their programming. ICEs give instant test-and-modify turnaround, but can cost thousands of dollars.

OTP ROM

One-Time Programmable ROM moves the job of programming from the factory to the customer. Using special programming hardware, you burn your code into the ROM at home, but only once. While MCUs are cheap, it still isn't very handy to throw one away each time you test a program.

EPROM

Erasable/Programmable ROM allows the user to erase the contents of the ROM and reprogram it. The erase process is done with an ultraviolet light and can take from 5 minutes to half an hour. With four or five chips, you can burn, test, and erase the chips in sequence with only a minimum of hassle.

EEPROM

Electrically-Erasable PROM is even easier to use, since programming and erasing are done at the same time, without using a slow UV erase cycle. EPROM and EEPROM don't allow an infinite number of reprogramming cycles, typically only around a thousand cycles (which is still three reprogrammings a day for a year).

Flash

Flash is a different flavor of EEPROM. EEPROM still needs to be programmed in a special device, and requires voltage levels above those used while running the ROM in the end circuit. Flash ROM can be programmed without taking it out of the circuit and with nothing more than logic-level voltages. It still has the 1,000 cycle lifetime limitation (though you can find Flash with greater durability), and writing to Flash is much slower than reading from it so you still can't use it like RAM.

SRAM

Static RAM isn't a read-only memory. It is a volatile memory like Dynamic RAM. The difference between SRAM and DRAM is that SRAM will hold its information as long as power is applied (and, with a small battery-backup, can hold its information for years). DRAM requires a constant refresh cycle to keep the information fresh, and specialized DRAM managers perform this refresh for desktop computers.

Many MCUs have SRAM internally for data storage, and some of them also allow you to add SRAM externally for more storage.

Special Options

There are many different features that can be built into an MCU. Here are a few of the standard ones.

UART

Universal Asynchronous Receiver and Transmitter. The UART understands how to communicate to the outside world through the standard forms of serial communication. This feature is especially useful if the MCU is going to communicate with an external device, such as an LCD display or a desktop computer. Chips that don't have an onboard UART can have one added externally through their SPI, or you can find or write software that emulates the UART.

A USART is a synchronous UART which has more operating limitations but is a lot faster.

SPI

Where the UART communicates with external systems, the Serial Peripheral Interface is used to communicate with other chips within the system. You can find a wide variety of chips with SPI ports: UARTs, A/D converters, even memory chips. Though not as fast as communicating through a parallel bus, SPI communication is simple to use and space efficient, using 3 or 4 wires instead of the full set of address and data lines.

I2C

The Inter-Integrated Circuit fulfills the same purpose as the SPI, with only two wires. The I2C protocol is better suited to network-like multiple-device communications, where the SPI is oriented towards one device talking to one other.

CAN

The Controller Area Network is a serial network communication scheme designed for use in automobiles. CAN fills the same needs as SPI and I2C, in a different market. This is a physically robust system, designed for harsh automotive environments.

I/O Ports

Digital I/O is important for any controller, and most MCUs try to maximize the number of I/O lines available. For each I/O line there must be a pin on the chip. An 8-pin chip is, by necessity, limited to just a few I/O pins. A 40-pin chip will normally have four 8-bit ports giving 32-bits of I/O.

I/O lines are often programmable. With programmable I/O, software determines whether the port is an input or an output by setting bits in a control register. Some MCUs give a mix of programmable and fixed I/O ports.

Analog Input

Since the real world doesn't always come to the MCU in tidy on/off signals, you can usually find a version of your MCU with analog inputs. These can be either analog-to-digital (A/D) converters, or analog comparators which take the analog signal and perform a threshold check, giving a digital value inside the chip.

When you can't find enough A/D inputs on an MCU for your application, you can always find external A/D chips to attach to the SPI port.

Timer/Counter

Timers run off the internal clock of the MCU. They can usually be scaled to run at different multiples of that clock (/2, /4, /16, /256). They can also be set to interrupt the chip when they overflow and roll around to zero. Timers are vital for any operation that outputs waveforms or signals at specific frequencies (such as PWM or R/C servo controls), or need to perform tasks at regular intervals.

Counters are like timers, but they operate from an external signal. The value of the counter can be read to see how many pulses have arrived at the counter's input port. The other features available on timers are often available on the counter as well. Most MCUs have one or more timer/counters.

Interrupts

An interrupt is a signal that stops the operation of the MCU and forces a special section of code to execute. Interrupts can come from outside the MCU, but there are internal interrupts as well, such as the UART character-received signal. The address of the interrupt handler code is stored in a special location called an interrupt vector. The interrupt vector is programmed into the MCU along with the rest of the application.

Interrupt handlers usually perform a short activity. When they are done, they turn interrupts back on (an interrupt usually turns itself off when it is activated), and return to the code that was interrupted. Care must be taken inside an interrupt handler to preserve the registers and flags of the MCU, so it doesn't adversely affect the execution of the interrupted code.

Without interrupts, the MCU spends most of its time watching and waiting at an input port for some intermittent signal to occur. With an interrupt, the MCU can perform more useful activities. When the signal occurs it is interrupted and can deal with it at that time. Interrupts can also be more reliable than polling, since even a very brief event will trigger the hardware interrupt but could be missed by software polling.

Watchdog Timer

Microcontrollers operate in the real world and, since perfection has eluded mankind, things go wrong. When an MCU is running the ignition of your car, it isn't terribly handy if it crashes and doesn't come back.

A watchdog timer, when activated, requires the application to notify the timer of its continuing good health at regular intervals. If the application misses its check-in time, the watchdog interrupts the system and allows it to restart or perform other error management activities.

PWM

Many MCUs have pulse width modulation capabilities built in, usually running off of one or more of the internal timers. Most MCUs do not have the radio control servo signal generation capability, but this can be emulated in software using the timers and software.

Basic Stamp ™

Basic Stamp is a registered trademark of Parallex, Inc. Basic Stamp and its variants are excellent products for people who want to use an MCU in their project, but don't want to worry about the difficulties of learning the details of the MCU.

These products run a program on the MCU that knows how to operate the MCU's various devices (like built-in A/D converters, serial communications, I/O ports, timers, etc.). This program can be controlled with simple and familiar Basic language commands.

Advantages of Stamp-type controllers include ease of use, increasing availability, and moderate cost for a full development system. The drawbacks include slow speed and limited program memory, though exceptions are appearing on the market. There are several books about the Basic Stamp with projects and tutorials.

8051 Family

The 8051 series of MCUs has been around for ages. It comes in many different varieties from many different companies, and has wide use in the automotive industry. If you find a CAN protocol on an MCU, it will probably be an 8051.

The 8051 also has the broad support of many commercial and shareware development tools.

68HCXX

These are Motorola's contributions to the MCU world. The 68HC11 has a huge following in the robotics community. This chip is the basis for several controller boards, kits, MIT's famous LEGO Brick, and so forth. This is also one of the few MCUs using Von Neuman architecture. The 68HC12 and 68HC16 add a number of powerful instructions to the HC11.

PIC™

PIC is a trademark of Microchip Technology, Inc. MicroChip's PIC processors are popular among hobbyists. From my personal experience my only complaint is the need for a UV eraser for the EPROM (though MicroChip is releasing new Flash versions). PICs come in many different sizes and with many different options. They are also easy to purchase, since they are well represented in the DigiKey catalog.

AVR™

AVR is a trademark of Atmel Corporation. Atmel has a series 8051 MCUs, as do most companies, but it's their AVR RISC processors that are of interest here.

The Atmel AVR MCU is much like the other MCUs, with a variety of sizes and options to choose from (including their new ATMega chips). Two features make it stand out in this list. Atmel has a Flash memory product line and their AVR MCUs all run off of a fast and easy to program Flash EEPROM. Atmel AVR chips are in-circuit programmable, which means you can hook a cable to your project and reprogram the microcontroller quickly and easily.

The Atmel AVR uses Harvard architecture with an instruction set designed to work well with high-level language compilers.

You can find several books and projects for the older, popular MCUs like the 8051, 68HC11, and PIC. Since the Atmel chips are new to the market, they don't have the following among hobbyists nor the range of tools and literature dedicated to them. This book intends to fill some of this gap. Not only does it use the Atmel MCU for the robot projects, the CD that comes with this book has a variety of useful software for programming the Atmel chips.

Project 7-1: Atmel MCU Base

There are two things you need for this project that must come from mail-order suppliers. The first is Atmel's AT8515 Microcontroller chip (and some of its supporting components), and the second is a device to program it. The MCU can be purchased from All-American, Arrow, Insight, and Marshall, among others (Atmel has a complete list of suppliers on their website). An Atmel programmer can be purchased from these same suppliers, or from SRS, who also carries a programmer in kit form. The rest of the parts can be found through normal channels. Note that if you plan on using Fuzbol (the language introduced in Chapter 10) you will need a special programmer that can communicate with the Fuzbol program loader. The SRS programmer does this, and the loader protocol is documented on the SRS web site for people who wish to create their own programmers. All parts for project 7-1 are listed in *Table 7-1*.

Qty	Part	Description
1	Capacitor, .1uf	Filter Capacitor
2	Capacitor, 18pf	Oscillator Capacitors
1	Connector, DIN-7	7-Pin Female DIN Connector
1	Crystal, 8MHz	Oscillator Crystal
1	IC, AT90S8515	Atmel MCU
1	LED	
1	Resistor, 1K Ohm	Resetup Pull-up
1	Resistor, 330 Ohm	
1	Switch, SPST Momentary	Reset Switch

Table 7-1. Project 7-1 parts list.

While you are waiting for these important components to arrive, you can read about the different Atmel MCUs and why these projects use the 8515.

Atmel has three lines of MCU: the AT89 series; an 8051-based 8-bit CISC MCU; the AT91 series ARM 32-bit RISC MCU; and the AT90 series 8-bit AVR RISC MCUs.

The AT89s are nice chips, though philosophically I prefer RISC to CISC. 8051 chips, however, are well documented in other books and have a broad base of support and information. Normally, when picking a chip for a project, this is a good thing.

The AT91s are also very nice, though they don't come in a package that makes them easy to use for the hobbyist (unless you like using TQFP) and they are overkill for the small pet robot. Not that overkill is bad, but they also cost more than the other chips. The AT90 AVRs are a small family, with only a few models compared to the 8051. AT90 MCUs have some very nice features and fulfill the three-bears criteria; they are not too small, not too large, but are just right.

Each of the AT90 series chips has in-circuit programming, an SPI port, a voltage comparator, and 32 general-purpose registers. Within the AT90 series, there are several different classes of chip:

AT90S1200 — 20-pin chip, 1 Timer/Counter (T/C), 1K Flash, 64 bytes EEPROM, 0 SRAM, 15 I/O lines.

AT90S2313 – like the AT90S1200, with 2 T/C, 2K Flash, 128 bytes EEPROM, 128 bytes SRAM, and a UART.

AT90S2323/AT90S2343 – 8-pin chips with only 3 I/O lines though twice the memory of the AT90S2313. The data sheets are not on the Atmel web site, and these chips seem designed more for commercial embedded processing.

AT90S4414 – 40-pin chip with 2 T/C, 4K Flash, 256 bytes EEPROM, 256 bytes SRAM, and a UART

AT90S8515 – the AT90S4414 with twice the memories.

ATMega103 – a new chip with massive memory (128K Flash, 4K EEPROM and SRAM) but without DIP packaging.

Though the 20-pin package of the AT90S1200 is much nicer to use on a small prototyping board, it has two drawbacks: no UART for easy communication to the PC or an external LCD serial display, and it has only 1/8 the memory of the 8515. The AT90S8515 costs about three times as much as the AT90S1200, but its $7.00 price tag shouldn't intimidate even the most timid hobbyist. We'll take the hassle of the larger and more expensive package for the benefits it can give us now and in more advanced projects.

Not listed is the new AT90S8535 MCU. This appears to be an attractive device with its built-in A/D converter. Note, however, that it does not have a provision for external SRAM. Other than these two variations, it is essentially an 8515 MCU.

Inside the 8515 MCU lies an entire world of possibilities, as seen in *Figure 7-1*. The external oscillator connection provides the clock necessary to run the processor. The four I/O ports provide an interface to the outside world. Port A and Port C are simple 8-bit configurable input or output ports. Part of Port B also doubles as the SPI port and/or an analog comparator. Part of Port D doubles as the UART interface.

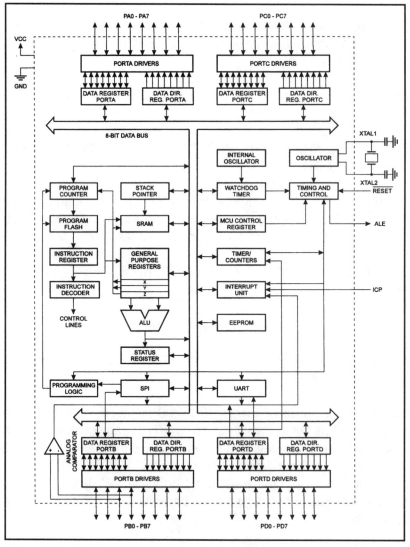

Figure 7-1. AT90S8515 Block Diagram. Copyright 1999, Atmel Corporation. Used with permission.

Internally there is the MCU processor which consists of the different memories, instruction decoder, ALU (arithmetic and logical unit, the part that does math and other math-like operations), various registers, timers and counters, and the SPI in-circuit programming system. All connection to the outside world is managed through the four I/O ports (plus a few special-purpose pins).

Adding an MCU to the robot is simplicity itself. The basic circuit consists of the MCU chip, some parts for the clock, and a connector for the programmer. An additional touch is the LED on Port-D Pin 4 to act as a heartbeat, to indicate the chip is programmed correctly. The only trick in building the circuit shown in *Figures 7-2* and *7-3* is to make sure the crystal and its capacitors are as close to the chip as possible, and to provide the cleanest possible power.

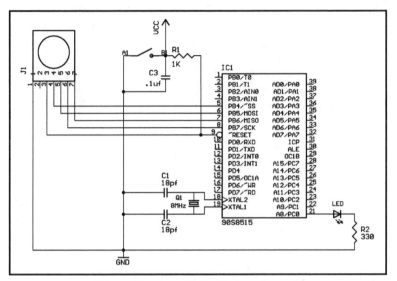

Figure 7-2. Atmel MCU base schematic

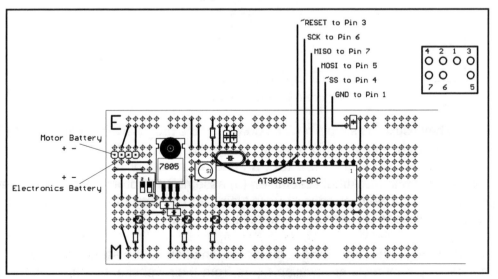

Figure 7-3. Atmel MCU parts layout.

The programmer used in this book is SRS's Atmel in-circuit serial programmer, shown in *Figure 7-4*. Most programmers, and most MCU chips, require you to remove the processor from the circuit and insert it into a socket on the programmer. An appealing feature of the Atmel MCU is their ability to be programmed in-circuit, without removing them from the project. You only need to find a place to attach the DIN-7 programming socket on the robot.

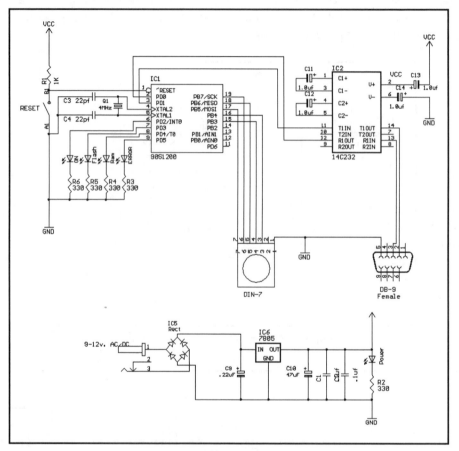

Figure 7-4. SRS Atmel programmer.

Once you have built this simple project, download the program alive.asm to the MCU. There are several steps in this process: (a) enter the program using a text editor (though *be sure* to save it as raw text, and not as an editor document), or use the editor supplied with the Atmel assembler, (b) assemble the program, (c) fix whatever typos or errors that were introduced while typing and go back to step (b) until it works. Finally, (d) download the program to the chip.

Steps (a) through (c) are covered in the documentation for the Atmel assembler. The Atmel development tools are available free from their website, and are also on the accompanying CD. It is also possible to skip all of the typing and copy the source code from the book's CD.

For step (d), programming the chip, there are two paths you can take. If you are using Atmel's programmer, or another out-of-circuit programmer, carefully remove the MCU from the robot and insert it into the programmer's socket. Following the instructions that come with the programmer, download the code. Then carefully remove the chip from the programmer and insert it back into the circuit. Power it up and hope the status light turns on.

Using an in-circuit programmer is even easier. Plug the cable from the programmer into the socket on the robot and turn on the robot's power. Following the instructions that come with the programmer, download the code to the chip. As soon as the download is complete, the chip should execute the code and, with any luck, the status light will turn on.

One of the challenges of using microcontrollers and computer software is the difficulty of finding out what went wrong when things don't work right. In the case of the first program downloaded to your first microcontroller, there are a host of things that can be wrong. The way to find the problem is by thoroughly and carefully checking everything; let no assumption go unchallenged. Is the power applied to the chip correctly? Is the crystal hooked up right, and oscillating? Is the ISP cable attached right? Is the status LED in backwards or correctly? Is the program entered exactly right? Is the programmer working?

Once the little light goes on, you know the MCU is alive. From there it is possible to add hardware and software to the system until it acts like a robot.

For the next section, the book takes a diversion through the world of machine language programming.

Machine Programming Basics

All programming language instructions can ultimately be traced down to a few fundamental concepts: Moving data from one location to another, performing simple operations on that data, testing data, and branching execution to a new program location.

Machine language (or, as the human-readable form is often called, assembly language) consists of exactly these types of low-level instructions.

Programming in assembly language is not difficult because the concepts are hard. It is difficult because the concepts are so simple (that is, primitive) that you, the programmer, have to be very clever in organizing these instructions to do any complex tasks.

To understand machine instructions, you have to understand the machine and how it is affected by the instructions. Human-readable instructions are converted to machine executable commands by way of an assembler. You need to understand a little bit about that process as well. All of this is easier to learn when you have a working example, which you can modify and observe the behavior caused by the modifications.

Some of the robot behaviors in this book are programmed in assembly language, and all of the examples and supporting codes can be found on the CD. A special robot control language called Fuzbol™ is explored in later chapters.

AT90S8515 Machine Architecture

Like most MCUs, the 8515 operates with a Harvard Architecture (as opposed to Von Neuman). Von Neuman machines have one area of working memory, which holds both machine instructions and data. Harvard machines have two areas of memory; one for machine instructions, and a separate area for data.

Both architecture forms have a set of special memory locations called registers, which hold temporary data inside the processor. Many operations work only with these internal registers. Registers in the 8515 MCU are also given regular memory addresses so they can also be manipulated like any other type of data memory.

A section of memory in the 8515 is reserved for hardware control. These are I/O control ports. The values in these registers control the behavior of the various hardware ports and peripherals built into the MCU.

Memory

Data Memory (SRAM)

A memory location is a piece of hardware that contains a set of values; bits which can be either set (on, or 1) or clear (off, or 0). These bits are collected into sets for easy access; four bits is a nibble, eight bits is a byte, and sixteen bits is a word. These sets of bits can be thought of as a string of zeros and ones, which is awkward. These bits value can be manipulated as a hexadecimal or decimal value. Hexadecimal is a base-16 counting system, and is a "natural" system for machine language. Base-10 is the decimal counting system we are familiar with in our daily lives.

A nibble can have any of 16 unique bit settings as shown in *Table 7-2*. In the "Hex" column you may notice that there aren't enough decimal digits to represent all of the bit patterns, so by convention you use the first six letters of the alphabet. If you work with machine language much, especially when doing bit-level operations, it is valuable to memorize the nibble conversion table. It is even better to understand it.

Bits	Hex	Dec
0000	0	0
0001	1	1
0010	2	2
0011	3	3
0100	4	4
0101	5	5
0110	6	6
0111	7	7
1000	8	8
1001	9	9
1010	A	10
1011	B	11
1100	C	12
1101	D	13
1110	E	14
1111	F	15

Table 7-2.
Nibble
conversion.

Each bit in a nibble represents a value. The first bit (bit 0) has the value of ($2^0=1$). Bit-1 has the value of (2^1)=2, bit-2 is (2^2)=4, and bit-3 has the value of (2^3)=8. To calculate the decimal value for a binary nibble, add up the values for each bit that is set (ignoring cleared bit values). Practice this and compare the results with the table. A byte has 256 unique bit settings. You can calculate the decimal value of a byte by adding up the power of two bit values like with the nibble (the highest bit, bit-7, has a value of (2^7)=128). In hexadecimal, however, a byte is simply two nibbles. Split the 8-bit byte into two 4-bit nibbles and convert the binary code for each nibble, which is simple. If you need a decimal value, it is easy to calculate from the nibbles; multiply the decimal value of the high nibble by 16 and add in the value of the low nibble:

> Byte: 01101100
> Nibbles: 0110 and 1100
> Hex: 6C (0110 = 6 and 1100 = C)
> Decimal: (6 x 16) + 12 = 108

Bytes can have values between 0 and 255 (if they represent positive values). In order to represent negative values, special consideration is given to the high bit (bit-7). When you subtract 1 from 0000.0001 you get 0000.0000. Subtract 1 again, and you get 1111.1111 (0xFF) which is considered to be -1. Subtract 1 again to get 1111.1110 (0xFE) which is -2, and so forth. When you reach 1000.0000 (0x80) you are at the value of -128. Subtract one again and you get 0111.1111 (0x7F) which is the value +127. This bizarre wrap-around also works when adding. In general, if the high bit is set, a value may be considered negative, and if the high bit is clear, the number is definitely positive.

Words can have 65,536 unique bit settings. Fortunately, a word is simply two bytes, or four nibbles.

In addition to having a value and bit pattern, each memory entry is associated with a unique identifier, its address. Think of a giant array of post-office boxes; each box has its own number, and each box holds a single value.

The 8515 has 512 data-memory locations laid out as shown in *Figure 7-5*, each of which contains an 8-bit number. The address of the first location is Hex 0060 (often written 0x0060 or $0060; the special characters before the number tell us it is in base-16 and not base-10). The address of the last location is 0x025f. Normally, memory addresses start at zero. In the 8515 the first 32 addresses are to be used to access the internal registers, and the next 64 addresses are used to access the I/O ports. These registers are said to be memory mapped, because they are mapped to (associated with) memory locations even though they are not really "memory" themselves.

Information in Static RAM is lost when power is removed from the computer, so this memory is only useful for temporary storage of information.

Program Memory (Flash)

Program memory is not only a different type of memory (Flash versus Static RAM), it resides in a separate address space. Instructions to access data memory cannot access program memory, and vice versa.

The 8515 has 4,096 (4 x 1024, or 4K) words of 16-bit program memory. Program memory starts at 0x000 and goes though 0xFFF. The data sheets says the 8515 has 8K of Flash; it means 8K by 8-bits, but it is actually used in 16-bit chunks.

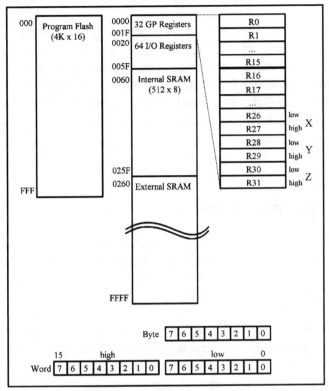

Figure 7-5. AT90S8515 memory layout.

In today's world of multi-megabyte computer memory, four thousand instructions don't seem like much. However, when you are coding each instruction by hand it can seem like a lot.

Stack

The stack is used by certain special instructions (PUSH, POP, and the various CALL and RET instructions). The stack is actually a section of data memory, starting at some high address (which you must specify in your program) and growing down into low memory. Most data memory is referenced by a specific address. Stack data is manipulated from the top of the stack, wherever that may be at the time.

A stack is usually described as being like the spring-loaded plate holders from a buffet. When you want to get a plate, you remove it from the top of the pile (stack). When plates are put back, they are added back at the top of the stack.

In a computer stack, when values are pushed onto the stack they are placed at the current stack position in memory. This top-of-stack position is then automatically moved down one address. When values are popped off of the stack, the top value is returned and the top of the stack is automatically moved up one address.

The stack is an important part of interrupt processing and the calling of procedures. During an interrupt or procedure call, the current execution address is pushed onto the stack. When the called code is done, that address is popped off of the stack so execution can continue where it left off. By using a stack instead of a fixed register or memory location to store the return address, it is possible for a subroutine to call another subroutine, and so on until all of the stack memory is exhausted.

Registers

General Purpose Registers

The 8515 has 32 general purpose (gp) registers named R0, R1, to R31. Having many registers is handy since almost all significant operations happen to or between registers. Each register is eight-bits wide, and is mapped to data memory so that R0 is at address 0x0000, R1 is at 0x0001, up to R31 at 0x001f.

The gp registers are divided into two classes: R0 through R15, and R16 through R31. The instructions that operate on a register, plus an immediate (constant) value (SBCI, SUBI, CPI, ANDI, ORI, and LDI), can only operate on registers R16 through R31. Three special 16-bit registers (X, Y, and Z) are used for advanced memory addressing modes. Each of these registers is constructed from 2 of the regular 8-bit registers; the X-register's low-byte is R26 and the X high-byte is R27. Y is R28 and R29, and Z is R30 and R31.

Status Register

The status register (SREG) is a special internal register that records the result of some instructions, and can be used to direct the operation of others. Each of the 8 bits in the SREG has a special meaning. These bits are tested, set, and cleared using special instructions. It is also mapped to the I/O register at address 0x003f and can be manipulated with generic instructions as well.

A summary of the status register bits is shown in *Table 7-3*. For more details on how the SREG is affected by operations, you need to refer to the documentation for the AVR, the instruction set. Each instruction has unique behavior with respect to the SREG flags.

Bit	Name	Description
0	C	Carry – If certain arithmetic operations overflow or underflow, it affects the carry flag.
1	Z	Zero – Set if the result of the last operation resulted in a zero value.
2	N	Negative – Set if the result of the last operation resulted in a negative value.
3	V	Overflow – Two's complement overflow.
4	S	Sign – A composite flag, the XOR between N and V.
5	H	Half-Carry – Set if there was a carry from bit-3 in the last operation.
6	T	Bit Copy Storage – Used by BST and BLD
7	I	Global Interrupt Enable – Set this bit to activate interrupts.

Table 7-3. Status register.

I/O Registers

The internal functionality of the 8515 MCU is exposed through the 64 I/O registers, also known as the I/O ports. Some of the general purpose registers (such as SREG and the stack pointer) are duplicated in the I/O register space while also being accessed through special instructions.

The I/O registers are accessed through the special port instructions IN and OUT. They can also be manipulated by generic instructions through their memory addresses. When used with IN and OUT, the I/O register numbers begin at port number 0x00 and end with port number 0x3f. As memory addresses, the I/O registers begin at address 0x20 and end at address 0x5f. The Atmel documentation lists both addresses for the registers.

Given the number and complexity of the I/O registers and the intricacies of their use, they are not documented here. Complete details can be found in the 8515 data sheet from the Atmel website or on the book's CD. As the I/O registers are introduced in the projects, they are described in greater detail.

Program Execution

The microcontroller executes one instruction at a time. The question is, how does it know which instruction to execute? And how can you affect the flow of execution? There is a special internal register called the Program Counter (PC; actually two 8-bit registers combined to make a 16-bit address). The program counter indicates which instruction is executed next. During normal operation, the MCU loads the next instruction from program memory, advances the program counter, and executes whatever operation it is instructed to perform. This continues until something special occurs, the machine is turned off, or the world ends (whichever comes first).

Barring hardware failure, there are several things that affect the smooth flow of execution. Direct operations on the program counter register, branch instructions, call and return instructions, and interrupts.

Since the PC is mapped into I/O register address space, you can fiddle with it like any other register. This is not recommended.

Branch instructions are the correct way to alter the flow of execution. Conditional branches change the PC when a certain condition is met. Conditions include things like a particular arrangement of status bits, or the value of a given bit in a register. Unconditional branches always change the PC. Branching allows the program to skip over instructions or to loop back and repeat an operation. "Call" and "Return" instructions also affect the program counter, and are given special attention.

"Call" instructions (RCALL, ICALL) take the current program counter address and pushes it onto the stack, and then reloads the program counter with the specified address. At the end of a "call" instruction, a return instruction (RET) pops the address back on the stack back into the program counter. This return has the effect of continuing execution where it left off in the calling code. This process is known as calling a subroutine, and its ability to execute independent sections of codes is vital. By breaking a program into many small (simple, and easy to understand and debug) subroutines, a complex problem is transformed into many simple problems. It is the same as when you take a complex hardware problem and divide it into several simple sub-modules.

Interrupts are a hardware-driven feature that are similar to the Call/Return process. When interrupts are enabled (both globally and for the hardware feature in question), there will come a time when the hardware wants attention. When this happens, it triggers an interrupt. An interrupt is similar to a call; the program counter is saved to the stack, global interrupts are disabled (to prevent an interrupt from being interrupted), and the program counter is loaded with the address of the interrupt subroutine. When that subroutine is done, it executes the "interrupt return" instruction (RETI) which re-enables global interrupts and continues processing where it left off. While interrupts preserve the program counter, they don't preserve the status register or any other register. Care must be taken to preserve these so the interrupted code will continue execution correctly.

The address of an interrupt subroutine (also known as the interrupt handler) is specified in the program by setting the address of the interrupt vector. A vector is simply a memory location that specifies the address of another memory location. The interrupt vectors are held in the first 13 addresses of program memory. The 14th address, 0x00d, is where the program actually starts. On the 8515 MCU, the interrupt vectors don't hold the address of the code to execute, but a relative jump (RJMP) command to that code which is effectively the same thing.

When the MCU first receives power, it executes a "reset" interrupt. The "reset" vector then launches the program for the first time. The "reset" code sets up the rest of the hardware and software values that may be needed, and continues on into the main program.

Interrupts are important when you need to manage several hardware modules at the same time and be responsive to each of them. The interrupt handlers themselves should be short; short enough so that their processing doesn't interfere with other interrupts that may be waiting for their turn at bat.

Addressing Modes

Since most of the machine's instructions involve either moving or operating on data, it is important to be able to tell the machine exactly where that data is. The location of data is known as its address, and the different ways of specifying an address are known as the addressing modes.

The wild variety of addressing modes can be confusing, and even the standard diagrams used to explain them can be unclear. Don't worry too much about all of this.

You will normally only use a few addressing modes and you can learn from the examples later in the book.

Immediate

Immediate data is coded directly into the machine instruction. Since it is part of the instruction, it does not have an address of its own.

Register Direct, Single Register

The data for a register-direct operation is stored in one of the general purpose registers *(Figure 7-6)*. The register number R_D is coded into the instruction. R_D may be either a source or destination for the instruction, or the operation may work in place. Examples of register direct instructions include:

LDI R_D, K	Load register R_D with the constant K
PUSH R_D	Place the value in R_D onto the stack
POP R_D	Pop the top of the stack into register R_D

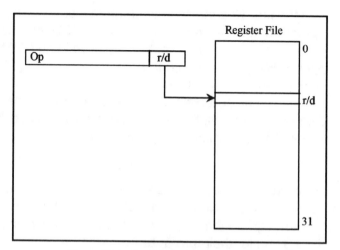

Figure 7-6. Register direct addressing.

Register Direct, Source and Destination

This addressing mode is used when two registers are required for the operation, R_R and R_D *(Figure 7-7)*. The result of the operation is stored in R_D. Each register's number is coded into the instruction. Examples include:

MOV R_D, R_R Move the value from R_R into R_D

SUB R_D, R_R Subtract R_R from R_D, and store the result in R_D

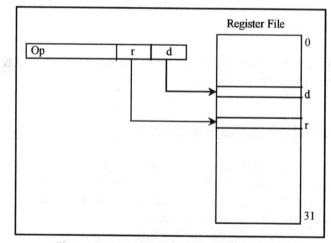

Figure 7-7. Two register direct addressing.

I/O Direct

Instructions that work with I/O ports code the port number directly in the instruction *(Figure 7-8)*. If the operation also works with a register, it will also be coded into the instruction like the register direct mode. Examples of I/O direct instructions include:

IN R_D, P   Input the value from I/O port P and place it into RD
SBI P, b   Set bit number b in I/O port P

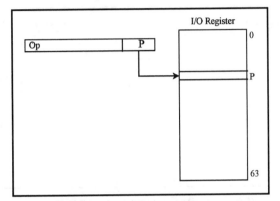

Figure 7-8. I/O direct addressing.

Data Direct

The data in this mode comes from the MCU's data memory. This memory may be from either of the register mappings, the 512 bytes of internal SRAM, or any optional external SRAM (for a total of 64K addressing space). The 16-bit data address is held in the 16 bits after the instruction, making this a 2-word operator. Examples of data direct *(Figure 7-9)* include:

LDS R_D, k   Place the contents of memory address k into register R_D

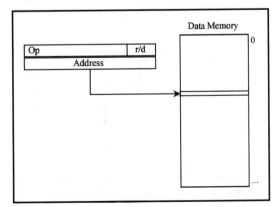

Figure 7-9. Data direct addressing.

Data Indirect

The address to data memory is kept in one of the special 16-bit registers (X, Y, or Z) as determined by the instruction. Examples of data indirect *(Figure 7-10)* instructions include:

LD R_D, X   Load R_D from the address named in the X register pair
ST Z, R_R Store the contents of R_R to the address named in the X register pair

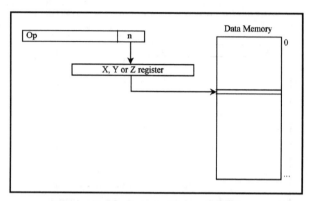

Figure 7-10. Data indirect addressing.

Constant Addressing

Constant addressing can also be called program indirect, and is only used by the LPM instruction *(Figure 7-11)*. The address in program memory is in the special Z register; the low bit indicates whether the low- byte (bit 0 = 0) or high-byte (bit 0 = 1) is to be loaded. LPM always stores its data into register R0.

LPM   Load R_0 from the address named in the Z register pair

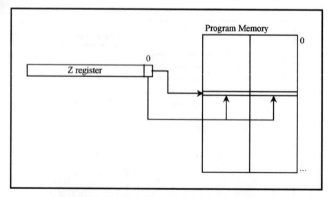

Figure 7-11. Constant indirect addressing.

Data Indirect with Displacement

Similar to Data Indirect, the address is held in the Y or Z register *(Figure 7-12)*. A constant offset is coded into the instruction, adjusting this address. This addressing mode is useful for high-level languages, where the register points to the base address, to a structure or object, and the offset shifts that address to a particular member of that structure. Examples of this addressing mode include:

STD Z+q, R_R Store register R_R into address Z plus offset q

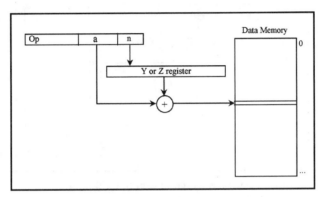

Figure 7-12. Data indirect with displacement.

Data Indirect with Pre-Decrement

Same behavior as Data Indirect, except that before the register value is used, its contents are decreased by one *(Figure 7-13)*. The changed address is stored back into the register before it is used. An example of this addressing mode includes:

LD R_D, -X   Decrement the X register pair, then load R_D from its address

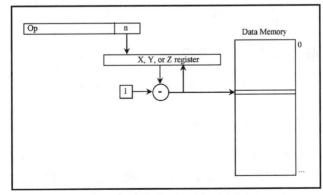

Figure 7-13. Data indirect with pre-decrement.

Data Indirect with Post-Increment

As in Data Indirect with pre-decrement, the register value is adjusted by one. In this case, the register value is increased by one and the new value is stored back into the register after it is used *(Figure 7-14)*. An example is:

ST X+, R_R Store R_R into the address X, then increment to the next address

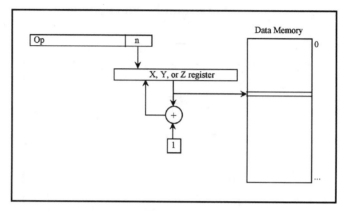

Figure 7-14. Data indirect with post-increment.

Indirect Program Addressing

Used by the IJMP and ICALL instructions, this mode transfers execution to the program address stored in the Z register *(Figure 7-15)*.

IJMP   Transfer execution to the program address held in the Z register pair

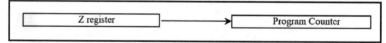

Figure 7-15. Indirect program addressing.

Relative Program Addressing

Used by the RJMP and RCALL instructions, this addressing mode adjusts the execution point by an offset of −2048 to +2047, specified in the instruction *(Figure 7-16)*. This mode is also used by the branch instructions. For example:

RJMP K   Change the program counter up or down K instructions
BRMI K   If the Minus flag is set, transfer execution up or down K instructions

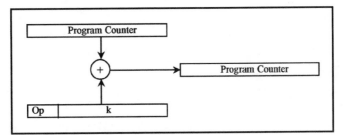

Figure 7-16. Relative program addressing.

Instruction Set and Development Tools

The details of the AVR-series instructions can be found in the manufacturer's data sheets. With a good understanding of the concepts in the previous sections, it should not be difficult to understand the operation of the various machine instructions. Another source of information is in the source code of example programs. The ultimate teacher is the MCU itself, or a good emulator.

The AVR development tools, such as the assembler and its editor, the programming tools, and debuggers are all documented by the manufacturer. Though it is outside the scope of this book to cover them in detail, take the time to review their documentation and to get comfortable with using these tools.

CHAPTER 8
MCU Senses

After a long chapter and lots of working with new and unfamiliar tools, the robot still just sits there, inert and lifeless. It is time to attach the new brain to some basic senses and muscles.

Program 8-1: Pulse

Mcu_heart.asm is a variation on the "alive" program used in Chapter 7; it adds a pulse to the indicator light. This pulse fills the same role as the timer-based heartbeat in earlier chapters. It regulates the responses of the robot. It also provides an opportunity to experiment with interrupts and timers.

The hardware portion of this project was built in Project 7-1, Atmel MCU Base. Though mcu_heart.asm simply causes the status LED to blink, it demonstrates the use of interrupt driven code, uses one of the timers, and it begins to control the world through the digital I/O ports.

This is an extremely simple program, so it is all in one file and it doesn't use subroutines. For this program, every section is described in detail below.

```
;================================================================
;
;        Robotic Heartbeat
;
;        Demonstrates:
;                Timer
;                Interrupts
;                Digital output
;
;================================================================
.include "8515def.inc"
```

The top comment block describes the contents and purpose of the file. After that, it includes a generic file which contains a variety of symbols used to support this particular chip.

```
;=======================================================================
;
; Put these into the upper-bank registers, so we can LDI them
;
.DEF Temp            = R16
.DEF BeatNum         = R17
```

The next section defines the two registers used in this application, both of which are in the upper register bank to make it possible to set their values with the LDI instruction. Temp is a throwaway, used for moving data in and out of the I/O ports. BeatNum is a counter used by the heartbeat, and counts down the number of interrupts left until the heartbeat actually toggles.

```
;=======================================================================
;
.EQU PORTC_HEART     = 0          ; Heartbeat flag Port C Bit 0
.EQU BEAT_PER_INT    = 24         ; Roughly 5 beats per second
```

Next are the symbolic constants. These numbers are used later in the program. Constants are given descriptive names to improve the documentation value of the code ("BEAT_PER_INT" conveys more information in the code than simply "24"). They are named at the top of the code to make it easy to change the behavior of things without having to read through everything. Want 10 beats per second? Change BEAT_PER_INT to 12.

```
;=======================================================================
;
.CSEG
.ORG 0x000

        rjmp    RESET           ; Power-on Reset
        rjmp    RESET           ; IRQ0
        rjmp    RESET           ; IRQ1
        rjmp    RESET           ; Timer1 Capture
        rjmp    RESET           ; Timer1 Compare A
        rjmp    RESET           ; Timer1 Compare B
        rjmp    RESET           ; Timer1 Overflow
        rjmp    Heartbeat       ; Timer0 Overflow
        rjmp    RESET           ; SPI Transfer Complete
        rjmp    RESET           ; UART RX Complete
        rjmp    RESET           ; UDR Empty
        rjmp    RESET           ; UART TX Complete
        rjmp    RESET           ; Analog Comparator
```

The CSEG (code segment) directive tells the assembler that we are writing code now. ORG gives the origin, or starting address, of that code. The first few addresses are then reserved for the interrupt table.

The interrupt table is padded with calls to RESET, with the addition of a HeartBeat vector on Timer 0.

```
;======================================================================
;
.ORG 0x0d

RESET:
        ; ——————————————————--
        ; Initialize the Stack Pointer!
        ;
        ldi     Temp,   low(RAMEND)
        out     SPL,    Temp    ;init Stack Pointer
        ldi     Temp,   high(RAMEND)
        out     SPH,    Temp

        ; ——————————————————-
        ; Indicator light on Port C
        ;
        sbi     DDRC, PORTC_HEART              ; Set Data Direction to OUTPUT
        sbi     PORTC, PORTC_HEART             ; Start with Heartbeat ON

        ; ——————————————————-
        ; Setup the heartbeat timer
        ;

        ldi     Temp, 4
        out     TCCR0, Temp                    ; Pre-scale Timer0 by 256

        ldi     Temp, 2                        ; TOIE0=1 (timer 0 interrupt on)
        out     TIMSK, Temp

        ldi     BeatNum, BEAT_PER_INT

        ; ——————————————————-
        sei                                    ; Activate global interrupts
        rjmp    MainLoop                       ; Go to the main program
```

Down in the working code, the RESET function (a) places the stack at the end of the SRAM using a constant defined in the 8515 include file, (b) sets the Port C Bit 0 to output mode and then turns on the light, (c) activates the timer and interrupts, then (d) jumps to the main program. The final jump is unnecessary, since MainLoop is actually the next instruction. I include the jump for clarity, and because if MainLoop is moved to the end of the program (for example) everything will still work correctly.

There are two timers in the 8515. Timer-0 has an 8-bit counter and very basic functionality, and Timer-1 has a 16-bit counter and many other features. The timers operate from some multiple of the system clock, or from an external source as an event counter. Each time a timer gets a pulse from the clock, it increments its internal counter by 1. When that counter overflows from 0x00, the overflow flag is set, and if it is enabled, a hardware interrupt is triggered.

The heartbeat uses timer 0. There are three mechanisms available to control the frequency of the timer overflow and interrupt. First, you select the frequency of the clock that drives the timer. In this case, it is the system clock divided by 256 giving 8,000,000/256 = 31,250 ticks per second.

Second, you load a value into the timer's counter (available through I/O Port TCNT0) causing it to overflow at any number of ticks from 1 to 256. When this method is used, it is important to reset the counter in the interrupt handler, otherwise it reverts to the default overflow of 256 ticks. The heartbeat program does not manipulate the counter, so it receives an interrupt once 31,250 / 256 = 122.07 times per second.

Third, the interrupt handler can count the number of times it has been called and only perform its action at some multiple of those calls. In the case of the heartbeat, it counts down from 24 to 0, and when it reaches 0 the heartbeat is executed. This toggles the heartbeat roughly 122 / 24 = 5 times per second.

```
;======================================================================
; Main Loop; does nothing
;
MainLoop:
        rjmp    MainLoop
```

The main loop of the program does nothing; it is a placeholder. All of the action occurs in the interrupts.

```
;======================================================================
;       Heartbeat
;
; Toggles the state of the heartbeat indicator
;
; This interrupt is called each time the heartbeat timer overflows.
; The system clock is 8,000,000Hz, with a pre-scale of 256 which
; gives 31,250 timer ticks per second.  The 8-bit counter will then
; overflow every 256 ticks, or about 122 times a second.
;
; We then manage this overflow with our own counter, giving us
; 5 pulses per second, or a pulse once every 24 interrupts.
;
Heartbeat:
        dec     BeatNum
        brne    hb_done                         ; If not zero, early exit

        ldi     BeatNum, BEAT_PER_INT ; Reset counter
;
        ; Toggle heartbeat by sensing current state and reversing it
        ;
        sbis    PORTC, PORTC_HEART              ; Skip next if heartbeat is set...
        rjmp hb_set                             ; clear, so set it...
hb_clear:
        cbi     PORTC, PORTC_HEART             ; Clear the heartbeat
        rjmp hb_done
```

```
hb_set:
        sbi      PORTC, PORTC_HEART              ; Set the heartbeat

hb_done:
        reti                                     ; Return from interrupt
```

The Heartbeat interrupt handler first decrements the beat counter, and when that reaches zero it tests the value of the heart light. The SBIS instruction is "Skip if Bit in I/O Register is Set", which means if the specified bit in the particular port is set, skip the next instruction. The "next" instruction is only executed if the bit is clear. In this case, it jumps to a section of code that sets the bit. If the bit was set, the jump is skipped and the second instruction clears the bit. Finally, everything funnels down to the "done" section, which returns from the interrupt.

This odd leap-frog structure for the "if-then-else" process is common, and you will find it in some form or other with many of the branching instructions.

Project 8-2: Touch

Qty	Part	Description
1	Capacitor, .1uf	Filter Capacitor
2	Capacitor, 18pf	Oscillator Capacitors
1	Connector, DIN-7	7-Pin Female DIN Connector
1	Crystal, 8MHz	Oscillator Crystal
1	IC, AT90S8515	Atmel MCU
3	LED	Indicators
1	Resistor, 1K Ohm	Reset Pull-up
3	Resistor, 330 Ohm	
2	Switch, Feeler	Feeler Switches
1	Switch, SPST Momentary	Reset Switch

Table 8-1. Project 8-2 parts list.

Before it starts dashing around the room, the robot needs its touch switches working again. The hardware aspect of this project is nothing more than adding two LEDs and two feeler switches to the MCU ports, as shown in *Figure 8-1*.

The feeler program mcu_feeler.asm tests to see if the feeler switches are working. When a feeler is touched, the corresponding LED should light up until the next heartbeat, when it is turned off. This demonstrates digital input and adds code to the main loop to monitor the input port. When something is checked repeatedly to see if it has changed, it is called polling. Polling and interrupts are the two basic methods of monitoring hardware.

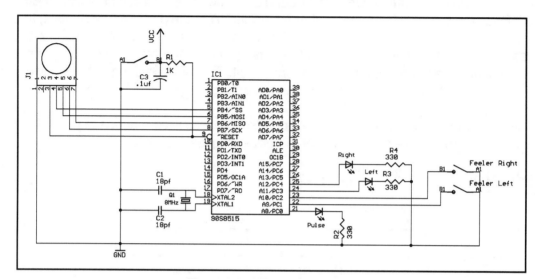

Figure 8-1. MCU with feelers.

```
;       ------------------------
; Feeler inputs on Port C
;
;       Both ports set to Inputs, with internal pull-up
;
cbi     DDRC, PORTC_LFI                 ; Input
sbi     PORTC, PORTC_LFI               ; .. with pullup

cbi     DDRC, PORTC_RFI
sbi     PORTC, PORTC_RFI
```

The digital input ports on the 8515 can be configured to be high-impedance (Hi-Z) or to pull-up with an internal resistor. When the port direction is configured for input (direction 0), the port's data value determines whether that input is Hi-Z (value 0) or pull-up (value 1). When they are Hi-Z, it is necessary to provide external pull-up on the feeler switch. To avoid this extra hardware the software sets the two input bits for internal pull-up. Because of this pull-up, the default state of the input pins is on (1). When the switch is touched it is turned off (0).

```
MainLoop:
        sbis    PINC, PORTC_LFI        ; Skip if left feeler is set (not touched)
        sbi     PORTC, PORTC_LFO       ; touched, so set the light
```

```
        sbis    PINC, PORTC_RFI         ; Right feeler touched?
        sbi     PORTC, PORTC_RFO        ; yes, set light
        rjmp    MainLoop
```

The polling code in the main loop is constantly checking the feelers, and as soon as one is switched the corresponding light is activated.

```
hb_clear:
        cbi     PORTC, PORTC_HEART      ; Clear the heartbeat

        sbic    PINC, PORTC_LFI         ; Skip if left feeler is clear (touched)
        cbi     PORTC, PORTC_LFO        ; Clear left feeler output

        sbic    PINC, PORTC_RFI         ; Right feeler free?
        cbi     PORTC, PORTC_RFO        ; Clear right feeler output

        rjmp    hb_done
```

The interrupt handler has additional code to turn off the feeler lights when the heartbeat goes off. Since the feelers can be activated at any time during the heartbeat cycle, the light will remain on a random length of time from not at all, to the length of one heartbeat cycle.

Project 8-3: Legs

The AT90S8515 MCU provides convenient dual PWM generation, so we will hook up that feature to the motor drivers and extend the previous program. The hardware changes are nothing more than hooking the motor driver to the appropriate MCU ports, as shown in *Figure 8-2*.

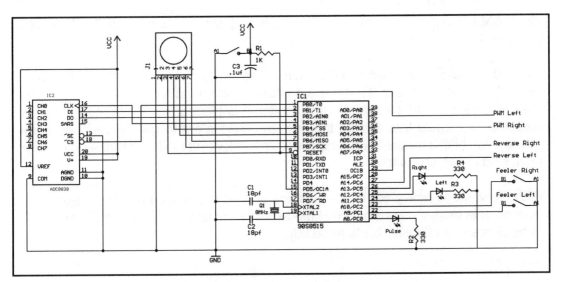

Figure 8-2. MCU motor hookups.

The motor program mcu_motor.asm adds pulse-width modulation control to drive the two "legs" of the robot. Its behavior demonstrates state-machines, a useful programming paradigm.

```
.EQU BEAT_PER_INT      = 61    ; Changed to roughly 2 beats per second
```

The old heartbeat at 5 pulses per second turns out to be too fast, so a simple adjustment of BEAT_PER_INT changes it to 2 pulses per second. Look at the program mcu_motor.asm and get a feel for the various parts. The significant code modules are explained in more detail below.

PWM Control

```
; ——————————————————-
; Setup the dual PWM timers
;
sbi     DDRD, 5                        ; Force the PD5 to be output
cbi     PORTD, 5                       ; to enable it as a PWM port

ldi     Temp, 0xa1
out     TCCR1A, Temp                   ; Set PWM mode on both counters

ldi     Temp, 0x02
out     TCCR1B, Temp                   ; Pre-scale by 8 on 8MHz clock

ldi     Temp, 0x00
out     OCR1AH, Temp                   ; Clear the high-bytes of the compare
out     OCR1BH, Temp                   ; register; only using 8-bit PWM

ldi     Temp, MOTOR_OFF                ; Start with both motors off...
out     LEFT_MOTOR, Temp
out     RIGHT_MOTOR, Temp
```

The PWM setup is somewhat intricate, as it is only one of several possible functions for Timer 1.

First, Port D pin 5 (also known as OC1A) must be set as an output for the PWM signal to reach the motor drivers. OC1B is a dedicated pin on the MCU, so it doesn't need special code to set its direction. Then the timer is placed into PWM mode and the prescaler (a module that divides the system clock) is set to give a reasonable frequency. Too high a frequency runs the motors weakly and too low a frequency makes them sing or, worse, twitch.

With PWM mode activated, starting values of zero are placed into the 16-bit comparator. This project uses 8-bit PWM mode, so technically the high byte doesn't need to be cleared, but it never hurts to be tidy.

To get an asymmetrical PWM pattern, the counter behaves in an interesting way. Starting at 0, the counter register increments once each cycle until it reaches a maximum value (which depends on the PWM mode; 0xff for 8-bit, 0x1ff in 9-bit, and of course 0x3ff for 10-bits). Once at the maximum, it then decrements back down to zero and starts the cycle over again.

In this particular PWM mode, when the incrementing counter reaches the value specified by the comparator, the PWM output port is cleared to zero. When the counter reaches its maximum value, it begins decrementing. When the decrementing counter reaches the comparator's value, the PWM output port is set to one. When the counter reaches zero, it begins incrementing again. When the comparator's value is zero, the output is always zero. When the comparator's value is one, the output is set for two cycles; from one down to zero and from zero back up to one (and the output is clear for all other cycles). When the comparator's value is the maximum of 255 the output is always set. This is illustrated in *Figure 8-3.*

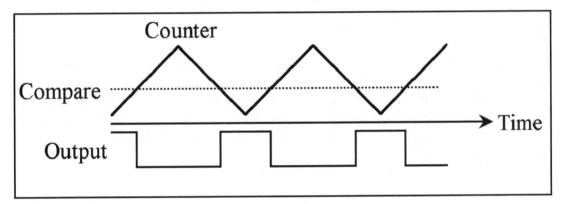

Figure 8-3. MCU pulse width modulation.

Feelers

```
MainLoop:
        ldi             FeelerOn, 0x00          ; Initialize this to zero
        rcall           Feelers                 ; get initial state of the feelers



;=======================================================================
;       Feelers
;
; Poll the feelers and update the status port and internal flag
;
Feelers:
        sbic    PINC, PORTC_LFI         ; Touched?
        rjmp    feel_right              ; No, check right
        sbi     PORTC, PORTC_LFO        ; Yes, so set the light
        sbr     FeelerOn, TOUCH_LEFT
```

```
feel_right:
        sbic    PINC, PORTC_RFI         ; Touched?
        rjmp    feel_done               ; No, done
        sbi     PORTC, PORTC_RFO        ; Yes, set light
        sbr     FeelerOn, TOUCH_RIGHT

feel_done:
        ret
```

The feeler polling code has been moved out of the main loop and placed into its own subroutine. This routine is then called throughout the state machine, to keep the status of the feelers up to date as the system executes.

The feeler subroutine manages the status lights, and has code to update an internal register value to mirror the state of the lights. This register is easier to test for the different combinations of feeler states than the individual bits on Port C, and is used throughout the state machine to control the transitions between states. The HeartBeat interrupt handler is updated to manage the FeelerOn register.

Behavior State Machine

State machines are useful for managing behaviors. This program's state machine is illustrated in *Figure 8-4*. State machines consist of three parts: states, transitions between states, and actions that occur during the transitions. The behavior coded in this project is crude yet sufficient to illustrate these concepts. Only the code for the "forward" state is listed here. Refer to the full listing for more details.

```
; ──────── Go forward, enter the forward state
forward:
        cbi             PORTC, PORTC_LREV        ; Signal the motors to go
forwards
        cbi             PORTC, PORTC_RREV

        ldi             Temp, MOTOR_MED
        out             LEFT_MOTOR, Temp
        out             RIGHT_MOTOR, Temp
state_fw:
        cpi             FeelerOn, TOUCH_RIGHT
        breq    turn_left
        cpi             FeelerOn, TOUCH_LEFT
        breq    turn_right
        cpi             FeelerOn, TOUCH_BOTH
        breq    stop_reverse
        ; TOUCH_NONE
        rcall   Feelers
        rjmp    state_fw
```

In the *Figure 8-4*, each state is a named circle. In the sample program, each state is identified implicitly based on which section of code is currently executing (such as the state_forward block shown here). A transition from one state to another is drawn as a directed arc between two states. This transition occurs only when a particular event triggers it. If there are no events, or an event does not affect the current state, no arc is drawn. The event that triggers a transition is written next to the arc in the diagram. In the sample program, the state code constantly polls to see if an important event has occurred and branches to the next state when one does.

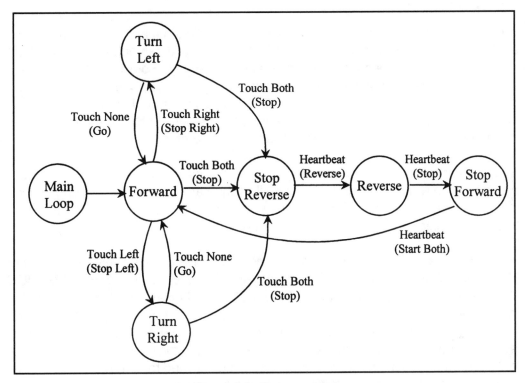

Figure 8-4. State machine.

The actions that occur during the transition between states is also written next to the arc in the diagram, in parenthesis. These actions do the work for the state machine, and occur before the destination state is entered.

For a machine as simple as the one in the example, this hard-coded form is acceptable. For advanced machines, it is easier to code each state and the possible transitions in the form of a table. This can be a single table where each entry lists a current state, an event for that state, a destination state if that event occurs, and the address of a function to execute when making the transition. Another form would be to have separate tables, one for each state listing the events and transitions for that state. Either way, a generic state machine interpreter could perform all of the event polling, state management, and action execution. Any change in the machine requires only a change in the table(s).

Of course, writing such a state system in assembly language is more work than it is worth unless you plan on using this technology a lot, or will be creating some very complex machines.

Debugging

Programs don't work the first time. On a microcontroller, in machine language, it can be very difficult to tell what is going on. With hardware you can poke at the wires with a voltmeter or an oscilloscope and see what is happening. For software you have to take extra steps in the software itself to make the internal state visible for debugging. The Atmel development system comes with a debugger, but since this code is triggered by sparsely occurring external events it can be difficult to use the debugger or simulator effectively. Below are some tips to help find problems.

When a change or set of changes in the code do not work, stop and go back to a version that did work. Then make one small change at a time to this working version, testing after each change. When it stops working, you've found the problem. Think about the problem and try solving it a different way.

If it is hard to tell if the code is working or not, you have to expose some of its inner data to the outside world. Unless your project uses every single port on the MCU, these make perfect windows into the inner software. For example, the sample program for this project did not work when I wrote it (as expected). To get an idea of what the state machine was doing, I sent the event variable FeelerOn to Port A and then watched it change as I manually switched the feelers. It turned out that I was misusing the SBR instruction and when I fixed that, things started to make sense again. No matter what the problem, the first step in fixing it is to find a way to make it visible.

In addition to sending internal variables out to the ports, different sections of code can write distinctive bit patterns to the ports to identify where the program is currently executing. This is the purpose of the LED in the very first MCU project. When that light turned on, you knew the program was running since the default state of all set to ports on the MCU after a reset is high-impedance input.

The secret to assembly language programming is to break the problems down so that you can code them in very small steps, each of which can be verified as working before you go on to the next.

Find friends with similar arcane interests, or join a club, so you can have extra pairs of eyes to look at difficult problems. Sometimes the simple act of explaining your bug will make connections in your mind and you will solve it even before you have finished explaining! I have made more than one phone call which ended in the phrase, "Oh! I know what's wrong... thanks!" even before my colleague had a chance to say anything at all.

Last, and surely not least, sleep on a problem. It is always easier to do tedious, difficult work when you are rested and have a fresh perspective on things.

Project 8-4: Light

Qty	Part	Description
1	IC, ADC0838	Serial A/D Converter
2	Resistor, 1K Ohm	
2	Resistor, CdS Photo	Eyes

Table 8-2. Project 8-4 parts list.

Eyes are analog senses. They give us the world in many shades of gray. The 8515 does not have built-in analog-to-digital converters, so you have to add an external device to convert the world's indefinite values to the digital numbers required by the MCUs binary point of view. Build the simple circuit shown in *Figures 8-5* and *8-6*.

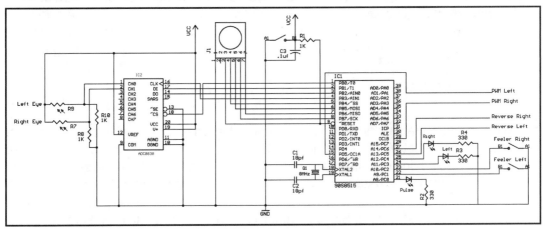

Figure 8-5. MCU with serial A/D.

Analog to Digital (A to D, or A/D) converters are made by most semiconductor manufacturers, and come in a wide variety of accuracy, speed, input channels, and interface options. The ADC0838 A/D is used here because of its low cost, easy availability, and multiple input channels.

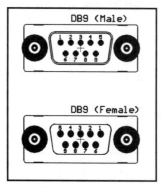

Figure 8-6. DB-9
connectors.

The ADC0838 interfaces to the MCU using a serial interface. AtoD converters (and many other peripheral chips) come in two basic interface types: parallel and serial. In a parallel interface, all of the data is read or written in one cycle using eight (or more) data lines plus a few control lines. Serial interfaces use less pins on both the MCU and the peripheral, requiring (typically) only a chip select, clock, data in, and data out lines. Some serial interfaces use only two wires. There are many flavors of serial interface, and the AT90S8515 has a very nice SPI serial interface built in. However, the ADC0838 does not use the SPI protocol but one just different enough to make SPI hard to use. What this chip does do, however, is provide an opportunity to show how to "bit bang" a serial interface. Mcu_eyes.asm demonstrates how to implement this serial interface in software.

```
; Analog-to-Digital Port B bits
.EQU PORTB_SELECT    = 0
.EQU PORTB_MOSI      = 1
.EQU PORTB_MISO      = 2
.EQU PORTB_CLK       = 3
```

The control bits are similar between most serial interfaces, whether they are SPI or not. The chip select bit does not have to be driven by the MCU, but when it is used more than one device can share the data and clock pins. Only the currently selected device responds to the data lines, with the rest presenting a high-impedance.

The MOSI bit is "Master Out, Slave In" and it carries data from the master to the slave. In the case of an MCU communicating with peripherals, the MCU is the Master and the peripheral is the Slave. MISO, then, is "Master In, Slave Out" and carries data from the slave to the master. CLK is the clock. On each cycle of the clock, one bit is passed across MOSI and MISO. In some protocols, data is transferred both directions at once. For the ADC0838, the first five bits are sent down MOSI to the chip, and the next 16 bits are received on MISO.

```
;========================================================================
;          adConvert
; Analog-to-Digital conversion
;
; The calling code sets adData to the channel to convert (0..7).
; The converted value is in adData on exit
;
adConvert:
        cbi   PORTB, PORTB_CLK
        cbi   PORTB, PORTB_SELECT
        rcall   ad_wait
```

To initiate an analog conversion, the calling code places the analog channel code into the adData register and calls adConvert. The channel codes are a bit odd, and follow the pattern shown in *Table 8-3*. These codes are for single-ended conversions. A double-ended conversion uses one analog channel as the reference (relative ground), and a second channel as the input to convert, like a differential op-amp. For more information on this chip's capabilities, refer to its data sheet.

Channel	Code
0	0x00
1	0x04
2	0x01
3	0x05
4	0x02
5	0x06
6	0x03
7	0x07

Table 8-3. Channel codes.

Inside of adConvert, the channel code has other control information added to it. This control information specifies the start bit, and what type of conversion is to take place. Then the clock is forced low (just in case) and the chip is selected. Chip select on the ADC0838, like on many other chips, is active low. A value of zero activates the chip.

```
;          ————
; Send the command out
;
ad_send:
        ori             adData, 0x18    ; Start, single ended
        ldi             ad_bit, 8       ; Send all 8 bits; only need last 5 though
```

Only five bits are needed to start the conversion process; the first set bit the chip receives starts the chip listening for the other control bits. Once all five control bits have been sent the conversion starts. Knowing this, the ad_send procedure sends all eight bits in the byte, with the top bits set to zero.

```
ads_tst:
        lsl     adData
        brcc    ads_0
ads_1:
        sbi     PORTB, PORTB_MOSI
        rjmp    ads_clk
ads_0:
        cbi     PORTB, PORTB_MOSI

ads_clk:
        rcall   ad_clock

        dec     ad_bit
        brne    ads_tst
```

The act of sending data through a bit in the port is, in this case, destructive. The data is shifted left (each bit is placed into the position just to its left, with the left-most bit ejected into the carry flag), and the carry flag is tested to determine whether MOSI should be set or cleared. Once set correctly, the clock bit is toggled to alert the converter. Each cycle of the clock (inside the ad_clock procedure) is surrounded by a wait call. Some peripherals can take the data as fast as the MCU can feed it; this chip wants its data between 10,000 and 400,000 bits per second.

```
        ; ————————
        ; An extra clock cycle
        ;
        rcall   ad_clock
```

There are two problems with using the SPI port to communicate with the ADC0838. The first is that the 8038 needs an odd half-cycle "pause" after writing the command and before reading the data. The second problem is that the phase of the clock is different for commands and data. By bit-banging, we have total control over the clock and its relationship to the data.

```
        ; ————————
        ; Receive the data
        ;
ad_receive:
        ldi     ad_bit, 8
        clr     adData

adr_tst:
        rcall   ad_clock

        lsl     adData
        sbic    PINB, PORTB_MISO
```

```
sbr     adData, 0x01

dec     ad_bit
brne    adr_tst
```

Receiving data is like sending it, in reverse. The clock is pulsed to tell the ADC0838 to send a bit, and then that bit is checked. The adData register is shifted to the left, with bits inserted as needed.

The converter can send 16 bits of data to the MCU. The first byte is sent MSB first, meaning, the most significant bit (bit-7) is sent to the MCU first. The second byte is a mirror of this first byte, sent LSB first. It is only necessary to read one of these bytes, and this software chooses to read the first one. The second byte is lost forever when the chip is deselected.

```
ad_end:
        rcall   ad_wait
        sbi     PORTB, PORTB_SELECT
        rcall   ad_wait

        ret
```

Finally, the converter is deselected.

This sample program only uses two of the eight analog channels available on the converter. Channel-0 represents the value of the left eye, and channel-1 is the right eye. These eye values are used to control the speed of the motors when the state machine is in "forward" mode (see the code for exact details). By adjusting how these values are scaled, and which motor they control, it is possible to easily recreate (in software) all of the hard-wired behaviors built in the earlier chapters.

Project 8-5: Interfacing to Displays and Computers

Qty	Part	Description
4	Capacitor, 1.0uf	
1	Connector, DB9	RS-232 Male DB-9 Connector
1	IC, 14C232	RS-232 Level Converter

Table 8-4. Project 8-5 parts list.

If you have survived the programming examples from the previous chapters, you are acutely aware of the need to see what is going on inside the MCU. You also know how difficult it can be to debug code with only a few binary outputs to guide the way.

For debugging, and for robot applications where full mobility is not required (or where your robot is large enough to hold an entire PC), it is convenient to interface to a desktop computer through its standard communications systems. There are two fundamental methods of transferring data to a PC, serial and parallel. In parallel communication, an entire byte or word is sent simultaneously with one wire per bit, plus some additional wires for control signals. Serial communication sends the byte or word one bit at a time down one wire, with control information preceding and succeeding the data. While the parallel port provides the fastest communication, the serial port is the simplest so that is what is described here.

If you choose to interface your robot to a PC, a new vista of opportunity opens up. Some of these avenues are explored here briefly once we get past the finer details of communicating to the PC.

RS-232 Serial Communications

The serial port on most computers follows the EIA/STIA-232-E specification, more commonly known as RS-232. New computers are adding USB ports, and USB may ultimately drive RS-232 out of existence, but that time is still far away.

RS-232 was defined in 1962, before personal computers were invented and even before TTL logic gates were invented. Its terms and features reflect its stodgy mainframe heritage. Formally known as an "Interface Between Data Terminal Equipment and Data Circuit-Termination Equipment Employing Serial Binary Data Exchange," the standard was designed to facilitate communication between peripheral devices (such as terminals and modems, the "Data Circuit-terminating Equipment" or DCE) and the computer itself (the "Data Terminal Equipment" or DTE).

Since TTL had not been invented, RS-232 does not conform to its 5-volt standard. It instead specifies voltage levels of +5 to +15-volts and −5 to −15-volts. In practice, systems typically use −12 and +12-volts, and these levels are generated by a special RS-232 interface chip. RS-232 is also active-low, so the −12-volt signal is a logical 1 (historically known as "mark") and the +12-volt signal is a logic 0 (or "space"). These, and other, interface coding issues are invisible to the RS-232 user, and are handled by a UART which formats the data with the various start/stop and parity information and manages any handshaking.

The full RS-232 specification calls for a 25-pin connector, and it uses most of those 25 lines for its various features. Today it is more common to see a 9-pin connector implementing a subset of the full specification. The male and female connectors, illustrated in *Figure 8-7*, are shown here looking at the pins and sockets of the connector. Note that when the connectors are nested together, pin 1 of the male connector goes into socket 1 of the female. The male connector is traditionally part of the computer (DTE) and the female connector is part of the peripheral (DCE). In a break from tradition, this book will sometimes refer to the DTE as the master and the DCE as the slave (in the SPI style). The signal descriptions for each pin are given in *Table 8-5*, from the point of view of the master.

Pin	Description
1	DCD (Data Carrier Detect). Generated by the slave when a suitable carrier signal is received by the slave.
2	RD (Received Data). Data line for information sent by the slave to the master.
3	TD (Transmitted Data). Data line for information sent by the master to the slave.
4	DTR (Data Terminal Ready). Generated by the master when it is ready to communicate with the slave.
5	Ground. Common ground between the master and slave (important for good communication).
6	DSR (Data Set Ready). Generated by the slave to indicate it is ready to communicate with the master.
7	RTS (Request To Send). Generated by the master when it is ready to actually transmit data to the slave.
8	CTS (Clear To Send). Generated by the slave in response to RTS, indicating the peripheral device is ready to receive data.
9	RI (Ring Indicator). For modems, set to on when the modem is receiving a ring signal from the phone line.

Table 8-5. DB9 pin description.

Due to the Host/Peripheral (or Master/Slave) configuration of RS-232, there are several different cables that can be used to connect devices to each other. In the simplest form, there is one master device with a male connector and one slave device with a female connector. The cable between them is a simple pass-through cable – for example, pin 2 of the male connector is passed straight through to pin 2 of the female connector. There are cases, however, where the cable will cross-over pins 2 and 3 to enable two master devices to communicate without colliding. All of the projects in this book use a common straight-through cable.

Simple RS-232 communication uses five lines of the 9-pin connector (TD, RD, RTS, CTS, and Ground). Trivial RS-232 communications disable the handshaking and use only TD, RD, and Ground. Handshaking can usually be disabled in software, in the serial port configuration. Failing that, the DTR, DSR, and CTS can be connected to each other and the master will then handshake with itself. Though handshaking increases the reliability of the connection, it is not strictly necessary and is left off for these projects.

Interfacing to a Serial LCD

The RS-232 interface for the AT90S8515's UART is shown in *Figure 8-7* and *Figure 8-8*. The DB-9 connector is configured to be a master (DTE) and it expects the other end to be a slave peripheral (DCE). For debugging and general communication for my robots, I use a serial LCD display such as the BPP-420L (four lines by 20 characters) or the really nice BGX-128L (128 by 64 graphics, plus various text support) from Scott Edwards Electronics (http://www.seetron.com/).

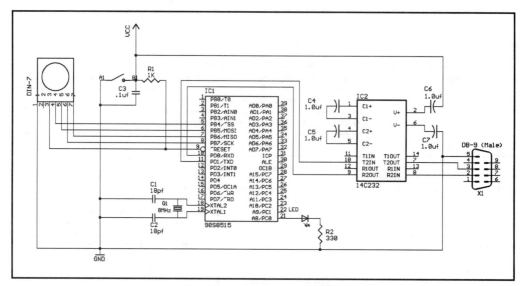

Figure 8-7. MCU with RS-232.

The serial interface is used in later projects, after Fuzbol is introduced. Fuzbol directly supports serial output using the communications settings of 9600 baud, eight bits, no parity, one stop bit (perfect for the Scott Edwards serial displays).

Interface to a PC Serial Port

To interface the robot's brain to a desktop computer, simply reverse pins 2 and 3 and use a female DB-9 connector. This converts the circuit so that it looks like a peripheral device and not a computer.

In order to see what is being sent to the PC, you have to run a terminal program to communicate with the computer's internal UART. Windows 95/98™ computers come with an accessory program "Hyper Terminal" which works just fine. Windows 95 and Windows 98 are trademarks of Microsoft Corp. Macintosh or other computer users will have to find a similar terminal program.

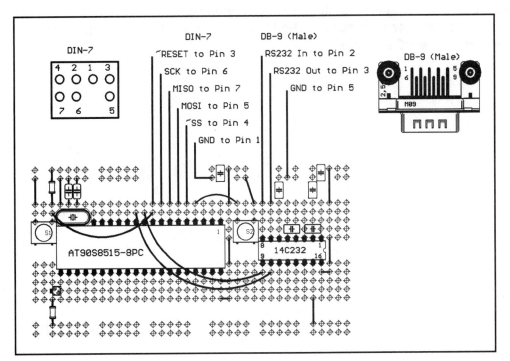

Figure 8-8. RS-232 parts layout.

After opening the Hyper-Terminal folder from the Start-Programs-Accessories menu, run (double-click) the Hypertrm.exe program. This will ask you for a name for the connection, and an icon to represent it. I call mine "Brains" and use the mad-scientist bubbling beaker icon. Next it asks for dialup information. Ignore that, and instead set the "Connect using" to "Direct to Com 1" (or whichever serial port you will be using). Finally, set the Bits per Second to 9600, Data bits to 8, Parity None, Stop 1, and Flow Control to None.

In addition to the debugging support possible with an LCD display or PC terminal program, having a wire to the desktop increases the possible intelligence of the robot a thousand or even million-fold, at the expense of full mobility. The PC can do the heavy thinking, and direct the (now semi-mobile) platform to carry out its instructions.

Wireless Communications

It is possible to have the best of both worlds – a fully autonomous mobile robot with the entire computing power of a modern desktop computer to drive it. The robot's local MCU can be programmed to handle only the most basic tasks, such as reading sensors, setting motor speeds, and so forth. It then communicates to the PC by way of a wireless RS232 modem and software on the PC can perform the difficult tasks of mapping the environment, planning activities, and generally being smart.

The PC directs the actions of the MCU by communicating in a pre-defined language through the wireless connection. The MCU responds with status reports to provide feedback for the PC's brain software.

CHAPTER 9
Igor, Fetch Me Some Brains

The control system makes a robot different from a mere machine. A really great control system can make your robot different from the many other robots out there. Up to this point the focus of this book has been on the hardware aspects of the robot, the body, and the physical brain. The behaviors have been simple, even crude. Their purpose was to demonstrate the basic functions of the hardware; simple reflexes to test the newly developed control pathways.

Robotic building, like so many other hardware-oriented endeavors, is no longer a simple question of silicon and wire. Microcontrollers are embedded in everything, and clever software is at least as important as clever hardware.

Layers of Intelligence

Though much of this chapter is theoretical (and cast in the dialect of the Mad Scientist, attempting to recreate life in his creations), the approach and ideas covered here apply to both the simple hobbyist platform and the technically sophisticated research vehicle. The technical literature is bursting with ideas and techniques for simulated intelligence, many of which can be traced back to a biological inspiration.

In the same way that different hardware systems in a robot build upon each another, the components of intelligence are created in layers. It takes several layers before a robot begins to develop interesting behaviors. The lowest level layer of intelligence is so primitive that it may be hard to think of it as a part of intelligence. This lowest layer is the hardware control layer, with such pieces as the PWM control system. In biology, even the peripheral nervous system has reflexes and learned reactions.

Above the hardware control layer is a reflex and posture control layer. Under reflexive control, an organism (including robots) has certain rapid, automatic responses to important stimuli. These reflexes are basic survival mechanisms, and as such must have a rapid response, bypassing any slow thought processes. After the reflex has occurred, the brain can respond and react to, adjust, or refine the action in progress. Reflexes create automatic behavior in response to external stimuli. Postural control simplifies the interface between the brain and the body. Simple instructions from higher levels are interpreted and executed automatically, so the higher brain does not have to worry about the fiddling details of making the body work correctly. Postural controls create behavior in response to abstract commands or stimulus from within the higher levels of the brain.

Additional layers of intelligence are not so well understood, but are easily recognized as the domain of the cognitive sciences – memory, planning, attention and intent, and so forth. A well rounded robot will have pieces of all of these, combined into a well-rounded whole.

This theory of layers of intelligence can be seen reflected in the Ontogeny of the human child, if you look at it under the right light and squint a little. Human Ontogeny (how behaviors change and increase in complexity over the development cycle) can provide a basis for the exploration of robot intelligence. On the mundane side, even the factory robots with their bland personalities are complex systems built up from many layers of control.

Birth to One Month: Adaptive Reflexes

The new child only shows basic responses to stimulus – reflexive and randomly triggered actions with little, if any, motivation behind them. Internally, it may be learning many little correlations of position and sensory input. Generating mental maps of the different positions its body can take, tracing paths in the hardware control layer's state space.

Feedback, both positive and negative, is provided by various hard-wired pleasure and pain stimuli. Systemic signals of comfort, discomfort, and the many other signals the body produces teach the brain what works and what does not. Perhaps the brain is learning in the form of Kohonen self-organizing maps, creating an internal symbolism to simplify and tokenize the complex interactions of mind and muscle. Or perhaps not.

This phase of intelligence relates to the training of the reflexive and postural control systems in the robot, correlating them with various internal and external stimuli. It creates the basis of the Postural Primitives explored in more detail at ftp://ftp.ai.mit.edu/pub/users/matt/sab.ps.gz. These reflexes are cyclic in nature. Sensory input affects the reflexive state, which directs the motor controller to perform an action, which affects and changes the sensory input... moderated by any hard-wired pleasure or pain responses.

One to Four Months: Circular Reactions

The basic reflexes explored in the first month are now chained together, creating repetitive motions. A sensory input (or directive signal from the still-unstructured higher brain) creates a motor reaction, which then leads to a second motor reaction. Visually-guided reaching begins to occur now, so the child is interacting with the world with more purpose and intent.

The brain is developing successful combinations of movements, or sequences of postures. At this point, the system may be calibrating vision with the tactile senses and motor control.

Four to Seven Months: Secondary Circular Reactions

Having just learned to sequence simple reflexive motions into a pleasant waving-about of the limbs, while watching and learning, the new person begins to notice the world's reaction to those wavings. Of course, it takes a while for a child to make a distinction between the internal world and the external world, so the correlation of internally directed actions is not much different from the correlation of externally affected activities. It all appears to be part of the same strange and mysterious domain of the newly developing mind.

In this new phase of development, the previously discovered actions are recreated in order to determine how they affect the external environment. Simple goal-directed behavior emerges. This begins the training of the plans-and-goals layer of intelligence; the map from the desire for a particular state to the chain of actions necessary to achieve it. The desired state may be a particular motion, position, or a simple modification of the world around us. Simple things, like *smack that bright thing* or *put my hand where I can see it*. The brain is learning to make use of cause-effect relationships.

Robotic intelligence begins to reach rough spots now. How are goals and desires structured, represented, or even identified for the robot? On a Mars mission, the roving robot has a set of hard-coded exploration tasks. These provide a solid foundation for the various intelligences and control systems to work from. For a hobby or pet robot... what makes it happy? Babies love pleasant cooing noises, bright colors, the face of its mother... food, warmth, gentle motion.

Side Note: What is interesting?

Intelligent systems are driven by things that interest it. Plants, for example, reach for light in a fairly mindless (though still efficient and complex) manner. Cats, though also phototropic, have a much richer repertoire of behavior than plants and are most noted for their curiosity – their pursuit of things that are interesting.

What makes something interesting?

Intelligent systems enjoy, and in fact may require, external stimulation. They prefer to dwell in those parts of the universal state-space that are interesting, rather than dull or frustrating. "Interesting" is a moving target which changes as the brain develops and learns. Things which are well-learned already, or are overly easy, are considered boring (with some notable exceptions – solitaire or knitting, for example, are considered more soothing than boring). Things which are far beyond the organism's current ability are considered frustrating (again, with exceptions – I usually find my current robotics project to be fascinating in spite of its extreme complexity). Only those tasks that are on a par with the brain's current development and educational level are really interesting. The *three bears* problem again; not too hard, and not too easy, but just right.

Seven to Nine Months: Coordination of Secondary Reactions

After spending a few months exploring various actions and reactions and how they affect the world and its position within it, it is time for the brain to take on a new task. Now the brain puts a new emphasis on intentionality and creativity, directing the newly developed reflexes and reactions towards achieving specific goals. The brain exhibits means-end behavior, including the use of intermediate actions to achieve the ultimate goal. For example, uncovering a hidden toy in order to play with it. Or perhaps more relevant, opening the fingers prior to grasping an object.

New actions are not being invented yet, but the brain explores the many uses (both familiar and novel) for the motions it has already learned. The brain begins to show signs of planning, and a stronger sense of intent. The developing human mind at this phase of development is already far beyond the state of the art in robotics by now, though there are parallels in various learning and planning systems.

Nine to 15 Months: Tertiary Circular Reactions

By now the brain has a fairly complete mental model of what its body can do, and what affect it has on the environment around it and its position within that environment. What new frontier can the brain open now? It is time to push the system to the limits!

Now is the time to perform a deep exploration of the state-space (the complete set of positions, actions, reactions, and correlations of the body and the world) of the immediate universe. This is an intimidating task, and one which the human child approaches with great enthusiasm. The system now pursues novelty for its own sake. If it's new or untried, it is interesting. Kittens seem to be born at this stage of development, with their driving sense of curiosity. The brain begins to direct the body to perform old actions in new contexts and to find new actions for old situations, performing experiments to see what happens. New and interesting outcomes are repeated and verified, not through hypothesis and intentional experiment, but by trial and error.

The brain is probably developing symbolic or representational models of its physical capabilities and the reactions of the world around it. Chaining cause and effect, exploring actions and reactions, and creating labels and ways of *thinking* about things and events. Exactly what happens in the human mind is still a mystery to us, but the clues it presents to the observer are still fascinating.

15 to 24 Months: Simulation of Events

Now the brain appears to have a functional, though not complete, internal representation of the world, and it can leave behind the old methods of physical trial-and-error. New behaviors are invented in the mind and they work correctly the first time when put into practice. This may mean that the rules and guidelines discovered by earlier trial-and-error are now represented internally. Perhaps they are held as some form of mental simulation which can be used to test new plans before committing to them physically.

During this exciting time, this representational model is being exercised, validated, and expanded. The same curiosity that drove the system at younger ages is still there, but now it is operating within the universe it has created inside the brain, where it can experiment safely and quickly. Language and symbolic communication are coming online now, and these tools are used to further expand the mental model of the world as well as to manipulate and affect that world in a very effective manner.

Summary

Each phase of development seems to mirror the one before. Each layer explores its state-space in a similar manner, but at a different level of complexity and abstraction always reaching forward to what is just out of reach.

At the lower levels of intelligence the hard-wired responses and reflexes from the physical body drive the learning process. Pain to be avoided, pleasure to be increased, basic needs to be fulfilled. As the layers build one upon the other, so does the abstraction and complexity of our needs until the need for mental stimulation and satisfaction motivates us as much as the earlier physical needs.

The Many Shades of Grey-matter

There are many different choices to make when designing a robot's brain. Is the brain hand-coded and hand-tuned, or does it create its own architecture through evolutionary methods? Are the brain's behaviors fixed reflexes, or is there learned behavior? What technology is used for the fundamental units of computation? Subroutines written in some computer language? Fuzzy rule systems, neural networks, or some combination?

The easiest way to develop simple behaviors is by writing them explicitly in computer code or with fuzzy rule sets. This is the approach taken in this book. The results are usually immediate and easily understood, which actually removes some of the mystery and excitement from things. It's like knowing the secret behind the magician's tricks.

In a learning system, the programmer lays the foundation and then turns it loose. The behaviors that develop, though guided by the programmer and the training data chosen, are still more than what was explicitly coded. There are two fundamental forms of learning systems, supervised and unsupervised.

In supervised learning, the engineer provides a comprehensive set of data—questions and their answers, or situations and correct responses—and the software then memorizes them. In neural systems, the software can also generalize the results and fill the gaps between the examples. Selecting good data sets, and choosing the learning parameters, is still something of an art though when this done well the results can be astounding.

Unsupervised learning turns the system loose in an environment and it then has to determine what is significant and what is not, without being explicitly told by the engineer. Some forms of unsupervised learning cluster input data, and categorize and classify information without being told in advance what the categories are. Other forms are given simple, hard-wired (or externally generated) "good" and "bad" feedback at relevant intervals. From this information, the learning system must then determine which actions and choices lead up to "bad" signals so they can be avoided, and which lead up to "good" feedback so they can be repeated. Reinforcement Learning is one such technique.

Hybrid techniques also exist, such as the neural networks that learn to drive cars. They "watch" a human driver manipulate the controls and learn what responses are necessary for different conditions.

Genetic programming is an extension of learning systems where the actual architecture of the brain is "learned." This "grown" architecture may be fixed in form, or it may learn from its environment. Evolutionary learning is a slow learning, where the more successful organisms or software systems are kept and the least successful ones are thrown away. Selection through random mutations and combinations between successful systems, new systems are created. Feedback from the environment or engineer slowly drives the architecture through state-space to an optimal (or near optimal) solution.

Learned or fixed, hand coded or genetically mutated, the programmer still needs to determine the units of computation for the system. Three popular forms of computation are structured programming, neural models, and fuzzy logic. Of course, there are other models as well as hybrids between these.

Structured Programming Models

This programming model is what most people use for robots today, though the balance may be shifting. When there is a problem to solve, the engineer or hobbyist sits down and figures out an explicit solution and then implements it in software. This solution works or it doesn't, and the programmer fiddles with it until it does or gives up in disgust. Such modules can be anything from inverse-kinematics for multi-axis factory robot arms, to path planning systems for maze-running mouse robots.

Neural Models

There are two levels to neural programming. The first is the simulation of individual neurons, which can be combined to create simple yet interesting reflexive behaviors. The second is the simulation of larger neural networks as a whole to create complex learning systems. There is a huge body of research behind the many neural models, and the field has developed many different system architectures, memory structures, and learning rules. The greatest drawback to neural models is that they are computationally expensive. It can take a long time and thousands of examples for them to learn their task. Of course, the human child needs several years of continuous learning before it is ready to launch out on its own, so everything is relative. New neural net hardware (such as the zero-instruction-set "ZISC" CPU) should improve the speed and power of neural computation in a few years.

Generic Neurons

In simplified terms, a neuron is a processing module that receives one or more positive (stimulating) and negative (inhibiting) input values. It adds up these inputs and if the result is greater than a given threshold value, the neuron "fires," creating a large output signal. *(See Figure 9-1)*

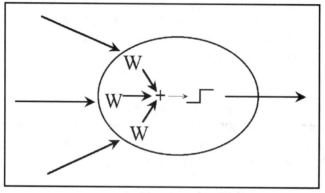

Figure 9-1. Neuron.

Each input may be weighted. An input weight scales the input signal with a priority from 0 (ignore) to 1 (vital), affecting its contribution to the calculation. Input values may be analog inputs from sensors, information from other neurons, or signals from other forms of software.

Adjusting these synaptic input weights is the primary method of training a neuron to respond differently to different patterns of input. In some models the weights can become negative, switching the meaning of an input from excitation to inhibition. In other models the sign of the synapse is fixed and the weights simply range from 0 through 1.

The summation process may simply be the addition of the inputs times their weights, or it can be a more complex function. There may also be an offset value added to the summation, so it doesn't start with zero when adding inputs. Offsets provide a base level of activation or inhibition. This offset can be used to tune a cell's sensitivity or create a neuron that always fires unless it is specifically inhibited. Changes in the cell's offset based on other activity in the model could be used as a learning function.

The threshold, the value which determines whether the sum of inputs turns the cell on or off, can also be fixed or variable. Thresholds may also change based on other cell activity, providing another avenue for learning.

The output of a biological neuron is a chain of pulses where not only their presence but their frequency is based on the level of stimulus the neuron is receiving. This gives the organism the strongest, least noise prone signal. Each pulse is always full strength and it is hard to miss. The sum of these pulses over time determines the recipient's activation level. In computer models, the neuron's output is typically a binary value of 0 or 1. If a more complex output is desired, neural models can generate pulse trains at various frequencies, or provide an analog-valued output.

Once a biological neuron fires, it has a refractory period where it is unable to fire again regardless of the stimulation. This refractory time sets the maximum frequency for that neuron.

Neural Networks

The neural network is more common than detailed models of individual neurons. Neural nets retain the concepts of inputs, weights, summation, and threshold, but the neural model has no refractory period and does not operate in the frequency domain.

The classic neural network uses the form of learning known as back-propagation. There will usually be three layers of neurons – an input layer, a middle "hidden" layer, and an output layer, as shown in *Figure 9-2*. Input signals propagate through the various connection weights and neuron thresholds until they reach the output. During learning this output is compared to the desired output, and any variations (errors) are passed backwards through the system. These error values are used to adjust the weights a little bit. The next time this same input is received it will generate an output closer to the desired result. When the network's trainer is satisfied with the network's behavior, training mode is turned off and the net is ready to perform in the real world.

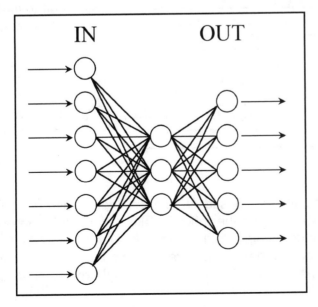

Figure 9-2.
Neural
network.

Reactive Control

Classical robot control, developed from the theories of early AI research, uses a very sensible computational model of intelligence. Sensors are first used to build a model of the immediate environment. A planning module determines the correct actions for the robot given the sensed environment and the robot's goals within it. Once this plan has been formulated, the robot executes it. This cycle repeats until the robot's goals have been achieved.

In practice, however, sensors can give unreliable data. Even with near perfect data, it is difficult to build accurate models of the environment, especially for hidden elements invisible to the sensors. Most environments are not static, but change randomly, complicating the problem. Planning is also a complex task and it can take a noticeable amount of time so the look, plan, and move cycle gives the robot rough motion.

In order to give more lifelike control to robots, a reactive control model was developed and refined over time; giving rise to the behavior-based control models. In behavior-based robotics, the robot's plan is implicit in the behaviors it is programmed to perform. The various behaviors are triggered and moderated by the current sensory input. The robot reacts to sensory changes with behavioral changes.

The best known form of behavior-based control is Rodney Brooks' Subsumption Architecture, an example of which is shown in *Figure 9-3*. There are four parts to the architecture: the sensors, the behaviors, the behavior selection (suppression) mechanism, and the hardware drivers.

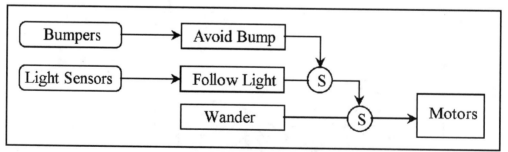

Figure 9-3. Subsumption architecture.

Sensors moderate the behaviors. Based on sensory input, each behavior determines the course of action it wants the robot to take, including no action. These actions are best represented as high-level commands such as "Turn Left," "Go Forward," and "No Action." The Wander behavior requires no inputs and always has an opinion on which direction the robot should go. Follow Light reads the values of the left and right photo sensor. If one or both of the sensors is below a preset value, this behavior has no action preference (No Action); otherwise it wants the robot to follow the light and sends the commands appropriate for that. Avoid Bump monitors the feeler switches, and, if they are touched, gives instructions to move away from the contact.

Each of these behaviors is evaluating their sensors and generating commands all of the time. In software, this can be implemented as a simple loop where at each pass through the loop, each behavior subroutine is called. This routine polls the appropriate sensors and sets the action code for that behavior. Since the MCU is fast, and each behavior is simple, it will act as if each behavior is running at the same time as the others.

```
Follow light:
    Get left eye value
    Get right eye value
    If left value and right value are too dim, "No Action"
    If left value is greater than right value, "Turn Left" otherwise "Turn Right"
```

Each behavior can be implemented in a way that makes sense for that behavior. Follow Light simply determines the relative light levels of the two sensors and sends the appropriate command to turn the robot towards the brightest one. Avoid Bump, however, is more complex and may be implemented as a state machine. It may need to stop the robot, back up for a certain amount of time, and then resume normal motion.

Once each behavior has determined their preferred course of action, it is necessary to choose which one to execute. In the subsumption architecture, only one behavior can be at the helm and which one wins is determined by the placement of suppression nodes. In this example, each suppression node has two inputs and one output. For example, the Avoid Bump behavior and the Follow Light behavior both feed one suppression node, and the output of that goes to the node fed by Wander. A suppression node makes a choice. It passes the commands from its left through unchanged unless it receives a command from above. The higher priority command from above suppresses the lower priority command. When Follow Light has nothing useful to say, wander gets free reign. When Follow Light sees light, however, it suppresses Wander and the light is followed instead. Of course, if Avoid Bump decides it wants control, its command moves down the line of suppression and is executed. One way of implementing this suppression is described here, where the highest-priority behavior is checked last and if it has a command, that command overwrites any previous commands:

```
Motor command is from Wander
If Follow Light is not "No Action" then motor command is from Follow Light
If Avoid Bump is not "No Action" then motor command is from Avoid Bump
Execute the motor command
```

The end result is a single command fed to the Motor control module. This command is then converted to PWM signals for the left and right motors, and the robot performs the action.

Behavior control methods, like subsumption, are an important piece of the robot behavior puzzle.

Fuzzy Logic

Where reactive control methods are described in contrast to computational control methods, Y is best described in contrast to classic Boolean logic. Fuzzy logic is a broad field with many different techniques available. What is described here is a simple form of fuzzy thinking.

Computer architecture is based on Boolean logic, which is a system of True and False, On and Off, 1 and 0. Everything is encoded with a distinct level of truth. When a question is asked, such as "Is the light bright?" the answer is "yes it is", or "no it is not" for any given definition of "bright."

Given one or more Boolean results derived from single questions, it is possible to combine these results using different operators to give more complex tests of truth such as "is the left light bright OR is the right light bright?" The three basic Boolean operators are NOT, AND, and OR. NOT implements the phrase "If the input is True then the output is False. If the input is False, the output is True". OR is: "If either input A is True OR input B is True, the output is True (otherwise it is False)". AND is: "If both A is True AND B is True, the output is True."

Boolean operators are described compactly in a table of inputs and outputs, called a truth-table. The truth table for three Boolean operators is shown in *Table 9-1*. The truth table for NOT shows that, for the input A, the output is its opposite. For OR and AND, there are two inputs – A and B, either of which may be True or False. Every possible combination of A and B is given with the results of the Boolean operation.

Lotfi Zadeh felt that the real world could not be represented so succinctly. For a given intensity of light, is it only Bright and Not Bright? If the light is Not Bright, but right on the edge of brightness, an infinitesimal change, a single photon's difference, will suddenly make it Bright. Or take a Bright light and remove one photon from it. Is it still bright? Take another photon.. and another.. until, suddenly, the remove of one critical photon makes it Not Bright. Boolean logic doesn't always model our intuitive understanding of the world.

NOT

A	not A
0	1
1	0

OR

A	B	A or B
0	0	0
0	1	1
1	0	1
1	1	1

Table 9-1. Boolean operations.

AND

A	B	A and B
0	0	0
0	1	0
1	0	0
1	1	1

Assume that illumination values from a sensor vary from 0 (total darkness) to 255 (total light), and that "bright" is considered to be an illumination value of 192 or greater. *Figure 9-4* illustrates the difference between Boolean and Fuzzy representations. These graphs are called membership diagrams, and they indicate the degree to which a statement is true. In the Boolean graph, the statement is either entirely false (degree 0) or entirely true (degree 1), and the transition occurs at precisely 192 illumination units. In the fuzzy graph, it is possible to be "almost" bright at a bit less than 192, "mostly" bright right around 192, and "entirely" bright a bit more than 192.

Fuzzification

A membership graph can take any shape at all, though for computational simplicity it is sufficient to use a triangle or trapezoid as shown in *Figure 9-5*. These membership graphs are defined by three values: the center, the width of the plateau on either side of the center (which may be zero for a triangle), and the overall width (or edge where membership is zero) on either side of the center.

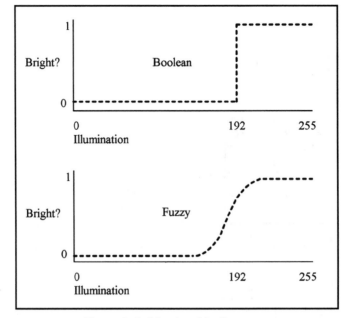

Figure 9-4. Membership diagrams.

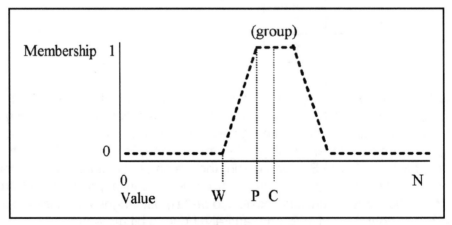

Figure 9-5. Fuzzy membership diagram.

The process of taking a value (such as a sensor reading) and determining its membership in a particular group is called fuzzification, as shown in *Figure 9-6*. The result is a value between 0 and 1 (inclusive) which gives the membership within the group. This membership value is the answer to the question being asked; the membership group is the question. The formula for calculating the membership weight given the center, plateau, width, and input value is:

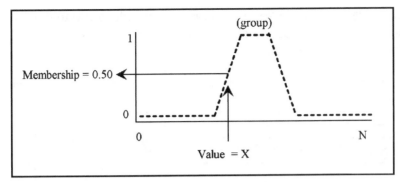

Figure 9-6. Fuzzify a value.

dist = | value – center |
slope =   width - plateau
weight =   0 : dist > width
1 : dist < plateau
1 – (dist – plateau) / slope : (dist <= width) and (dist >= plateau)

Fuzzy Rules

A question answered alone is not useful. Now that we know the extent the light is bright, what can be done with the knowledge? It is necessary to provide a "then" clause to the question, such as "If this condition is true, then perform that action." These questions and results make up the rules of a fuzzy system.

Fuzzy logic with a single question behaves almost the same as Boolean logic, so it is necessary to expand the system to two questions. Fuzzy logic provides a mechanism to smoothly shift between the answers to two or more questions. Let's look at a problem in subsumption control with two variable behaviors plus the root-level Wander. See *Figure 9-7.*

When nothing else is going on, this system moves towards the brightest light. If it hears any significant sound, it runs away from it. In a Boolean subsumption architecture, once the sound level reaches a given threshold, its behavior entirely suppresses the light following behavior to run away from the sound. In a fuzzy system, things can be quite different.

Both of the behavior modules generate motor commands all of the time. Follow Light generates commands to move towards the light. Avoid Sound moves away from sound. In addition to their commands, each module generates a weight value (a priority level, or behavioral imperative) that goes with the command. The weighted behavior rules read "To the extent that the light is bright, follow the light" and "To the extent that the sound is loud, run away from the sound.". The weights follow the fuzzification rules, and the closer the sound or light values are to the "perfect" condition of brightness or loudness, the closer the weight is to True (1). Wander has a constant low priority level constant.

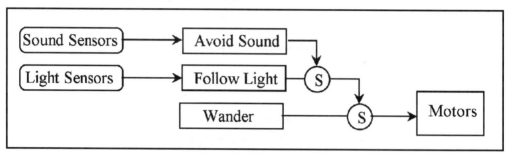

Figure 9-7. Fuzzy subsumption.

If the suppression nodes simply did a comparison of the weights, the resulting behavior would not be any different than the Boolean form. If the sound is louder than the light is bright, the avoid sound behavior takes full priority. If the light is brighter than the sound is loud, the robot follows the light instead. If both sound and light are at low levels (lower than the wandering weight constant), the robot wanders. Interesting, yet not using the abilities of fuzzy logic. The process of combining multiple weighted fuzzy results into a single crisp value is addressed by the defuzzification process.

Defuzzification

Motor control commands are ultimately reduced to speed values for the left and right motors. Lets assume that these motor values are what the behavior modules create. In addition to the speed values to guide the robot, each behavior is producing its priority value so that the command reads "to the extent that this command has a high priority, make the motor go this speed." With this information, it is possible to combine several different weighted commands to reach an optimal motor speed given the conditions.

This combining of several fuzzy values to reach a useable (crisp) value is called defuzzification. There are many different ways to defuzzify the results of a rule set, many of which involve taking the weighted centroid of various polygons which have been truncated by the fuzzy membership weights. If you want to know what that means, check the references for fuzzy logic; these system can get terribly complex.

There are good reasons to use the complicated defuzzification methods, and none of those reasons include simplicity. In a simple defuzzification scheme, the crisp result value is simply a weighted interpolation of each of the fuzzy value/weight combinations. This is called a point, or singleton defuzzification. In special cases it will fail dramatically. If one behavior wants to turn left and another wants to turn right with the same priority, the robot may stop dead, or worse, continue to drive straight into the obstacle. With careful planning these exceptions can be avoided.

Assume Avoid Sound wants to go a speed of 1 with a weight of 0.25 and Follow Light wants to go a speed of 6 with a weight of 0.75. We can tell that the light is brighter than the sound is loud (it has a greater weight), so the expectation is that the defuzzified speed will be closer to 6 than 1. The formula to calculate this defuzzification is:

$$((1 \times 0.25) + (6 \times 0.75)) / (0.25 + 0.75)$$
$$= (0.25 + 4.50) / 1.0$$
$$= 4.75$$

The generalized formula to find a crisp value from a set of fuzzy results is:

$$value_{crisp} = \sum(value^i_{fuzzy} \times weight^i) / \sum weight^i$$

Fuzzy System

In general, a fuzzy system looks like *Figure 9-8*. In the subsumption example, there were two rules each with one fuzzy range of interest (Bright and Loud). Their possibly conflicting results were defuzzified into one reasonable result. In typical fuzzy systems, the entire sensory input range is covered by overlapping membership functions so that for every possible input value there will be positive memberships in one or more ranges. There is also a rule for every possible condition, and these rules will then blend smoothly from one result to another as the inputs change.

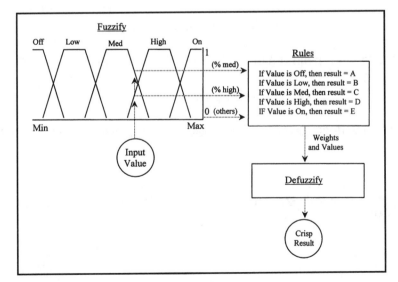

Figure 9-8. Fuzzy rule system.

There may be multiple fuzzy systems in operation at one time, each generating a crisp result from their sensory data inputs. These results could be cascaded as inputs into other fuzzy systems, and so forth.

Each aspect of the fuzzy system affects the end behavior. The shape, number, and placement of the membership functions determines how finely the rules can discriminate between different input values. The membership functions can be clustered around important input ranges, and narrower functions give a finer grain of control. Likewise, a single function can cover a broad range of values that are not important or don't affect the output in much detail.

The rules themselves can be more than simple if/then clauses; they can combine multiple fuzzy values from different inputs to generate their result. The defuzzification can be by the singleton method, or one of many other defuzzification methods.

Fuzzy Operators

Fuzzy rules can be more complex than the simple if/then tests used so far. Just as Boolean logic has operators to combine values, so does Fuzzy logic. Simple fuzzy analogs to AND, OR, and NOT are described next. There are actually many different ways to implement these, and other, fuzzy logic operators.

Fuzzy NOT simply inverts the fuzzy weight, as shown in *Figure 9-9.*

$$\text{NOT } w = 1\text{-}w$$

Fuzzy OR takes the maximum of two weights, shown in *Figure 9-10.*

$$w_A \text{ OR } w_B = \max(\ w_A,\ w_B\ )$$

Fuzzy AND takes the minimum of the two weights, as seen in *Figure 9-11.*

$$w_A \text{ AND } w_B = \min(\ w_A,\ w_B\ )$$

Where truth tables describe Boolean operations, membership graphs illustrate fuzzy operations.

In each of these cases, the fuzzy operators behave exactly like the Boolean operators for weights of 0 or 1. Fuzzy logic is an extension of Boolean logic, not necessarily a replacement.

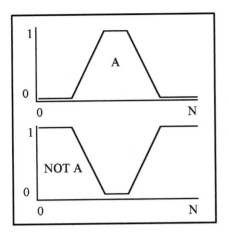

Figure 9-9. Fuzzy NOT A = (1-A)

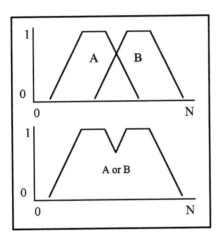

Figure 9-10. Fuzzy A OR B = max (A, B)

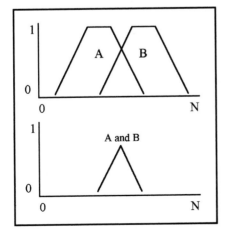

Figure 9-11. Fuzzy A AND B = min(A, B)

CHAPTER 10
Fuzbol

Fuzbol: Fuzzy Control Language

Performing even simple operations in assembly language can be a tedious and frustrating task. Most programmers prefer to use a higher-level language such as Basic or C and their variants. These languages simplify the task of managing memory and variables, and the instructions they give the programmer are more powerful. Each high-level instruction will incorporate many machine instructions. High-level languages allow the programmer to think in terms that are closer to the problem, and ignore irrelevant details. They are also slower than hand-coded machine language; in some cases, as much as 100 times slower.

Fuzbol is a new language the author has developed for his personal robotic control projects. It is an interpreted language, which means it does not create machine language instructions but special codes which are then executed by a special program, called an interpreter. The penalty in execution speed from using in interpreted language is compensated for by the increased ease in writing programs.

Fuzbol provides a fuzzy data type to manage fuzzy information, plus some special instructions to support fuzzification of data and fuzzy rule evaluation. Another feature of Fuzbol is how its structure is designed to support microcontroller hardware. Most other languages written for MCUs are simplified versions of standard Basic or C, which is a great benefit when it comes to learning how to use them. Fuzbol, though very similar to C in structure and syntax, is unique. The user will need to spend more time learning the language, yet the author's hope is that the special features of Fuzbol compensate for this effort.

The version of Fuzbol included with the book is a preliminary version. Regular updates and information can be found on the SRS web site. The appendix includes additional Fuzbol documentation.

Project 10-1: Using Fuzbol

Build the Atmel circuit with RS-232 system from Project 8-5, shown again in *Figure 10-1.* From the CD or off of the web, install the Fuzbol compiler onto your Windows '95 compatible computer. Attach an in-circuit programmer (ICP) between the computer and the SPI port of the chip. Finally, connect the RS-232 port to an ASCII display (either a second serial port on your computer, or an LCD display). This chapter assumes a Scott Edwards serial LCD is attached, though any serial display should work.

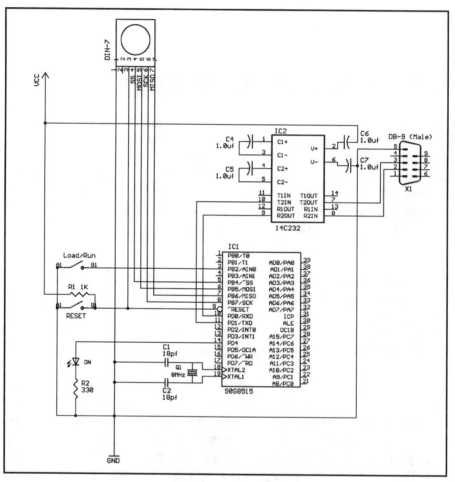

Figure 10-1. Fuzbol MCU.

Fuzbol Setup

Once you have assembled the necessary components, run Fuzbol on your PC. You should see the application as shown in *Figure 10-2.*

The first time you use Fuzbol, you need to tell it a little something about your environment. Starting at the bottom of the Fuzbol dialog, you need to select and configure the communications port used to talk with the ICP. The SRS ICP communicates at 19,200 bps, 8 data bits, 1 stop, no parity, and no flow control.

In the next box up, the fields labeled "DATA" and "CODE" indicate where Fuzbol should store data and program codes, on-chip or off-chip in external static RAM. The "External RAM" field indicates, in K-Bytes, how much total memory is available for use on the MCU system. To make sense of these fields, it is important to understand how Fuzbol and the 8515 work together to manage memory. See *Figure 10-3*.

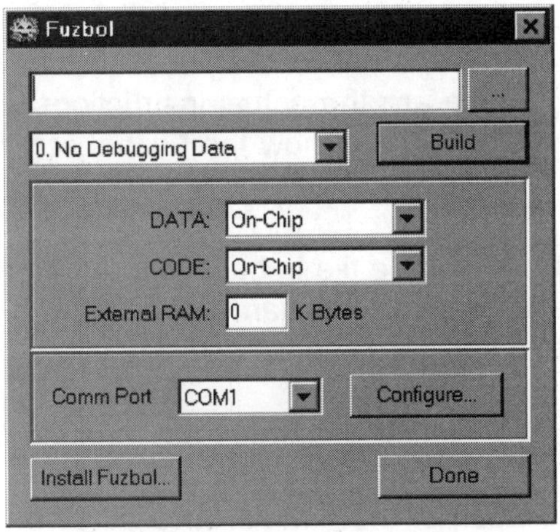

Figure 10-2. Fuzbol compiler.

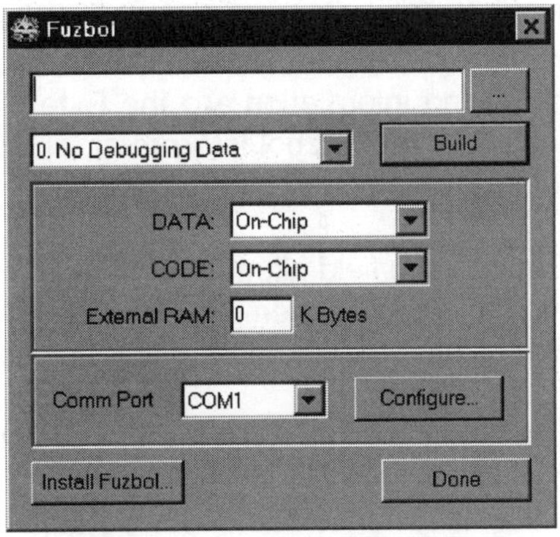

Figure 10-3. Fuzbol memory layout.

The stack always starts at the top of the internal RAM and grows down to the bottom of RAM. The stack is used whenever a local variable is declared, and whenever a function call is made. On-chip data begins at the bottom of the internal RAM and grows up towards the stack. The data section holds all object variables, and can be quite large in data-intensive programs. On-chip code begins immediately after any on-chip data and continues to grow up towards the stack. There is no protection mechanism to prevent the stack from growing into the bottom of code space, so be careful.

Either data or code may be stored in external RAM, if it exists. Doing this eliminates the possibility of a stack conflict, however, every access to off-chip memory costs an additional cycle of CPU time so there is a penalty in speed.

There is a limited amount of memory available inside the 8515 MCU (512 bytes, to be exact), so for complex programs it becomes necessary to add additional memory to the chip. The 8515 allows up to 64K total RAM, but when the external memory is activated it uses all of Ports A and C plus Port D bits 6 and 7 to operate. The next few projects rely entirely on the internal memory of the chip, so set both CODE and DATA to On-Chip, and the External RAM to 0.

All of these settings are recorded by Fuzbol in the registry, so you never have to reset them. At least, not until you change the hardware.

In the main area of the dialog, there is a choice of debugging information from None (0) to Way Too Much (3). This compiler output was designed to help debug Fuzbol during development, and is left in for the truly curious (or masochistic) user. The most interesting information are the Byte Codes. These are the actual instructions executed by the Fuzbol interpreter, and they give a sense of how the Fuzbol source code is transformed by the compiler.

The very top field allows you to select the Fuzbol source file to build and download into the system's memory. You can type in a full path and filename, or use the "..." browse button to the right of the field to locate a file. Don't select any files yet.

The "Build" button compiles the selected Fuzbol file. It generates any debugging information to a file called "fuzbol.dbg" in the Fuzbol program directory, and either indicates an error has occurred during compilation or presents a success screen. This screen indicates how much memory is used by the program and asks for permission to download the program to the MCU.

Install Fuzbol

Before any Fuzbol program can be downloaded and run on the MCU system, the Fuzbol interpreter must be installed on the chip. This section assumes that you are using the SRS ICP. If you use any other programmer, follow the instructions for that programmer to install the Fuzbol.hex file. This file is in the Intel standard format "Intellec 8". Note that, though any programmer can install Fuzbol onto the MCU, it takes a special programmer to download Fuzbol program codes into SRAM.

With everything hooked together and the MCU and ICP powered on, click the "Install Fuzbol..." button. The Fuzbol compiler then asks for the file to program into the MCU flash memory, and defaults to "Atmel\Fuzbol.hex." This should be correct, though if Fuzbol.hex does not show up in the File Open dialog, search around until it can be selected.

Once Fuzbol.hex is selected, click on "Open" and the process of sending the Fuzbol interpreter to the MCU begins. Once complete, the process will either give an error or a status display showing the progress of the download. The Fuzbol interpreter fills almost all of the onboard flash, so the download takes a few minutes.

Some popular errors in downloading include:

Connection timed out; problem with programmer or connection settings.

There was no communication between Fuzbol and the programmer. Make sure everything has power and the correct communications port is selected. Check the settings of that port, and check the cable connection between the PC and the programmer.

Garbled Connection; problem with programmer or connection.

The connection between Fuzbol and the ICP exists, but the data is getting damaged somewhere in transit. Make sure all of the connections are good, the MCU has stable power, and that the communication settings are correct.

Error retrieving...

During the startup of the ICP, Fuzbol asks it several questions about the programmer's type and various other things. If any of these questions are not answered by the programmer, then something is amiss. The programmer may be the wrong type or version to work with the Fuzbol loader.

ERROR VERIFYING DATA address...

Though the ICP is responding to the PC, it isn't having any luck with the MCU. Make sure the MCU has power and the SPI connection to the programmer is good. Ensure that the programmer can pull the MCU reset line to ground, since this is vital to the programming process.

Writing Programs

Fuzbol expects its source code to be in a clean, simple text file. Programs can be written using any editor that can create a text file, but such file formats as Microsoft Word's "doc" format have hidden codes and other invisible content that are not acceptable to the compiler.

Running Programs

After the Fuzbol interpreter is installed on the MCU, the chip is reset and prints the cheery message "Fuzbol v1.1" and "Loader:" to the RS-232. This means the program loader is active and waiting for a Fuzbol program.

Fuzbol programs are sent to the chip (by way of the ICP) by the Fuzbol compiler when the "Build" process is successful.

After a Fuzbol program has been successfully sent to RAM, Fuzbol displays "Done" after the loader prompt, and then the line "Fixup: Done" if the codes are good. Any variation from this pattern indicates an error.

Once a program is in memory and verified as good, hold the "run" switch (which lowers Port B Pin 2 to ground) and press reset. This restarts the Fuzbol interpreter in program-run mode, prints the message "Running..." to the RS-232, and executes any codes in memory. Once the program is finished, Fuzbol sends "Done" to the RS-232 and goes to sleep. This does not mean everything is finished executing. Any interrupt handlers are now activated, and may be triggered by the hardware.

Program 10-2: Fuzzy Heartbeat

The first Fuzbol program is the Heartbeat. Since Fuzbol allows either hardware polling or interrupts, but is not tolerant of mixed models, this program will be written as a timer interrupt in anticipation of future interrupt-driven extensions. The full code can be found on the CD as "Heartbeat.fuz". Note that all Fuzbol source files should have the "fuz" suffix, and they compile over to a "hex" suffix.

```
use "system_8515.fuz";
```

The system_8515 file defines the System object, which has named variables representing each control register and most of the significant bit names for the 8515 MCU. Most Fuzbol programs will need to include this System object.

```
define Heart
{
        const PULSE_CNT = 6;
        const ON = 0x10;

        count as integer;

        do Init();
        do Pulse() as interrupt 7;
}
```

The only object in this test program is the Heart object. Inside of the Heart are two constants, PULSE_CNT which determines how many interrupts occur before the LED toggles, and ON which defines the pin number the LED is attached to (the Pulse() method assumes the port is PORTD).

The object variable "count" keeps track of the number of times the Pulse() method is triggered.

Finally, Heart has two methods: Init() which turns on the timer and its interrupt system, and Pulse() which is marked as being attached to interrupt 7. Interrupt 7, as seen in the Fuzbol reference, is the Timer-0 overflow interrupt.

```
Heart:Init()
{
        count = PULSE_CNT;

        System:DDRD = System:PIN0_OUT;          // Output
        System:PORTD = System:PORTD & ~ON;      // Off

        System:TCNT0 = 0;                       // Init counter to 0
        System:TCCR0 = System:T0_CK1024;        // Clock step once every 1024 cycles
        System:TIMSK = System:TOIE0;            // Enable Timer-0 overflow interrupt
}
```

The first line of the Heart:Init() procedure sets the count variable to the delay multiple defined in the object definition. Note that symbols defined in an object, when used in some other part of the object, don't need that object's name to qualify them. For example, PULSE_CNT as used in Heart:Init() would need to be called Heart:PULSE_CNT if referenced outside of Heart.

The second section sets up System:PORTD, so that the appropriate pin is an output, though Port-D Pin-4 is already set up as an output by Fuzbol. Fuzbol uses this pin as an indicator. Note that Heart:Init() turns off the light.

The final section sets up the System Timer-0, and the overflow interrupt for it.

```
Heart:Pulse()
{
        count = count - 1;
        if (count <= 0)
        {
                count = PULSE_CNT;
                if (System:PORTD & ON)
                {
                        System:PORTD = System:PORTD & ~ON;
                }
                else
                {
                        System:PORTD = System:PORTD | ON;
                }
        }
}
```

The Heart:Pulse() interrupt handler is called every time Timer-0 overflows every 256 x 1,024 clock cycles, or about 61 times every two seconds.

The first thing Heart:Pulse() does is decrement the counter. If the counter has not reached zero, nothing else occurs and the MCU goes back to sleep.

Once the counter reaches zero, it is reset and then the System:PORTD heartbeat pin is tested. Depending on the result of this test, the LED is turned on or off.

That's it. Essentially the same code as used in mcu_heart.asm. The Fuzbol version will run about eight times slower, and it takes twice the number of bytes to represent. Its almost enough to make you wonder why anyone would want to use Fuzbol instead of assembly language! This question should answer itself as more complex programs are developed.

Project 10-3: Fuzzy Legs and Feelers

Recreate the robot with feelers, with the Fuzbol-enabled MCU as shown in *Figure 10-4*. The Fuzbol source is found in the "Legs.fuz" file.

This project incorporates the behavior from Heartbeat, and adds interrupt-driven feeler switches plus dual hardware PWM outputs. Sadly, Legs.fuz is about the most complex Fuzbol program that can fit into the on-chip RAM. The next project expands the microcontroller to a level where it can manage a far wider range of robotic systems.

```
define Robot
{
        beat as byte;                           // Heartbeat counter
                const PULSE_CNT = 10;
                const HEART_ON = 0x10;

        left_vel as byte;                       // Left motor velocity
        right_vel as byte;                      // Right motor velocity
                const VEL_ADD = 8;
                const VEL_MAX = 128;

                const LEFT_REV = 0x80;
                const RIGHT_REV = 0x40;

        do Init();                              // Initialize Heartbeat timer
        do Pulse() as interrupt 7;              // Interrupt handler on Timer-0
        do LeftFeeler() as interrupt 1;         // IRQ-0
        do RightFeeler() as interrupt 2;        // IRQ-1
}
```

The Robot object defines a heartbeat counter "beat," plus the constants used by the heartbeat subsystem, the left and right wheel velocity variables, various constants used by the motor drivers, and a variety of methods to implement the heartbeat and feeler interrupts.

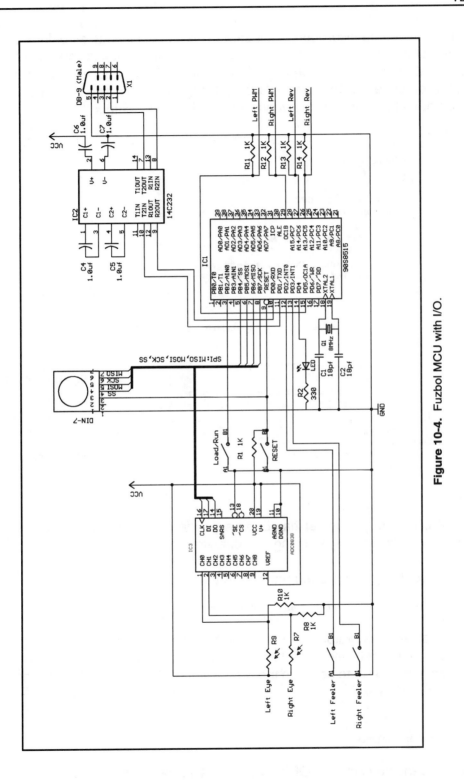

Figure 10-4. Fuzbol MCU with I/O.

The Init() procedure incorporates the same Heartbeat init code as before, and adds code to setup the feeler and PWM ports. Each feeler switch is wired to one of the external interrupt pins. These are set as inputs with pull-up enabled to minimize external components. The interrupts themselves are set to trigger an interrupt when the port goes from high to low, whenever a feeler is first bumped.

```
Robot:Pulse()
{
        beat = beat - 1;
        if (beat == 0)
        {
                beat = PULSE_CNT;
                System:PORTD = System:PORTD ^ HEART_ON;

                if (left_vel < VEL_MAX)
                {
                        System:OCR1AL = left_vel;
                        left_vel = left_vel + VEL_ADD;
                }
                else
                {
                        System:PORTC = System:PORTC & ~(LEFT_REV | RIGHT_REV);
                }

                if (right_vel < VEL_MAX)
                {
                        System:OCR1BL = right_vel;
                        right_vel = right_vel + VEL_ADD;
                }
                else
                {
                        System:PORTC = System:PORTC & ~(LEFT_REV | RIGHT_REV);
                }
        }
}
```

The Pulse() interrupt handler has been expanded from it's original heartbeat role, and now includes velocity ramping and direction control.

The heartbeat toggle has been reduced to the single statement:

```
System:PORTD = System:PORTD ^ HEART_ON;
```

The exclusive-OR operator "^" is useful for this type of two-phase cycling. It can also be used to toggle between two arbitrary values. For example, if you wanted to have a register or port alternate between the two values ALPHA and BETA, you can declare a third value TOGGLE:

```
TOGGLE = ALPHA ^ BETA;
```

Now, any variable exclusive-OR'ed with TOGGLE will switch from ALPHA to BETA and back:

```
my_var = ALPHA;
my_var = my_var ^ TOGGLE;                    // my_var = BETA
my_var = my_var ^ TOGGLE;                    // my_var = ALPHA again
```

At each heartbeat, the left and right velocity variables are incremented by a fixed amount until they reach the pre-defined maximum. Once this maximum is reached, both motors are then forced forward. This disables the reversing turn that is set up by each feeler interrupt.

```
Robot:LeftFeeler()
{
        left_vel = VEL_MAX/4;
        right_vel = VEL_MAX/2;
        System:PORTC = System:PORTC | LEFT_REV | RIGHT_REV;
}

Robot:RightFeeler()
{
        left_vel = VEL_MAX/2;
        right_vel = VEL_MAX/4;
        System:PORTC = System:PORTC | LEFT_REV | RIGHT_REV;
}
```

When a feeler's interrupt is triggered, the appropriate handler sets the motor velocities and directions for a broad, slow curve away from the contacted feeler. The Pulse() procedure will slowly increase these speeds until one of them reaches the maximum, and then put the robot back into forward motion.
Legs.fuz causes the robot to wander in roughly straight lines across the environment. It bumps into things and then backs away from them.

Advanced MCU

When your sole helper is an uneducated hunchback, you have to simply live with what brains he brings back from his midnight acquisition runs. If you end up with the deformed gray matter of a homicidal maniac, well, what can you do? Fortunately, the home roboticist has more control over their creation's processing units and we don't have to settle with inadequate brains. 512 bytes of RAM rapidly becomes insufficient and it becomes necessary to look for better hardware.

A number of MCU architectures allow for external memory, and the Atmel 8515 is one of them. By using all of Ports A and C, plus a few lines of Port D, the Atmel 8515 can be convinced to address up to 64K of external RAM. This is more than sufficient for most projects.

Project 10-4: MCU with External RAM

Qty.	Part	Description
9	Capacitor, .1uf	
1	Capacitor, .22uf	
4	Capacitor, 1.0uf	
2	Capacitor, 18pf	Oscillator
1	Capacitor, 47uf	
1	Connector, 2 x 16	2x16 32-pin Male or Female Header
4	Connector, 2 x 5	2x5 10-pin Male or Female Header
2	Connector, 2-pin	2-Pin Male Header
1	Connector, DB-9	RS-232 Male DB-9 Connector
1	Connector, DIN-7	7-Pin Female DIN Connector
1	Crystal, 8MHz	Oscillator
1	IC, 14C232	RS-232 Level Converter
1	IC, 62256	32K x 8 SRAM
1	IC, 74HC02	Quad NOR Gate
1	IC, 74HC138	Address Decoder
2	IC, 74HC573	Address Latch, Input Port
1	IC, 74HC574	Output Port
1	IC, 7805 or LM340	Voltage Regulator
1	IC, ADC0848	Parallel A/D Converter
1	IC, AT90S8515	AVR MCU
1	IC, Rectifier	Bridge Rectifier
2	LED	
1	Resistor, 1K Ohm	
2	Resistor, 330 Ohm	
2	Switch, SPST Momentary	Reset and Run Switches

Table 10-1. Project 10-4 parts list.

The bare-minimum circuit that demonstrates the use of external memory is shown in *Figure 10-5*. In the earlier MCU circuits, communication to the peripheral devices was managed through only a few wires. External memory as shown here is using a parallel data and address bus, which is fast but more complex. Too complex to be built on the average prototyping board. Once development has reached this stage, it is better to create a printed circuit board. There are a number of methods for creating boards at home, including transfer systems that allow you to print the layout on a laser printer. There are also a variety of services for printed circuit board prototyping, though they can be very expensive. It is sometimes possible to simply buy predesigned boards that fit your needs.

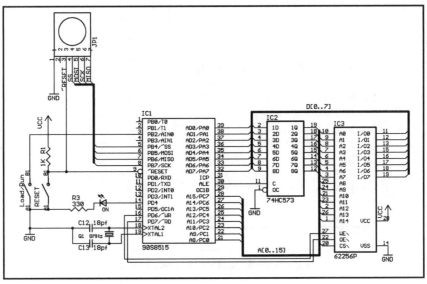

Figure 10-5. MCU with external SRAM.

It is not entirely necessary to provide external memory using a parallel bus. A number of systems implement interpreted environments like Fuzbol using serial memory chips; external RAM that is read and written through the SPI port. These are easier to build, but impose a significant slowdown on the language. It may be possible to convert the Fuzbol interpreter to use serial memory by rewriting the next_command macro and some of the branching functions.

The left-hand block of the schematic is the usual stuff. An SPI port for programming, a couple of switches for Reset and Load/Run control, the Heartbeat indicator, and the MCU itself.

In order to use the least number of pins to support external RAM, the MCU uses Port-A for double duty. During one part of the read/write process Port A holds the address to be accessed, and then during the later part it holds the data for that address. IC2, the 75HC573, is used to latch the address so that, in conjunction with the other address bits at Port C, the full address is always visible on the bus.

The process of accessing external memory is a complex interplay of signals, shown in *Figure 10-6*.

A single access to external RAM may take three to four clock cycles, depending on whether the MCU is configured to have a wait state or not. T3 is the "wait" cycle, which may be necessary for slower peripherals. For example, the wait state is enabled in software when the A/D converter shown in *Figure 10-7* is accessed.

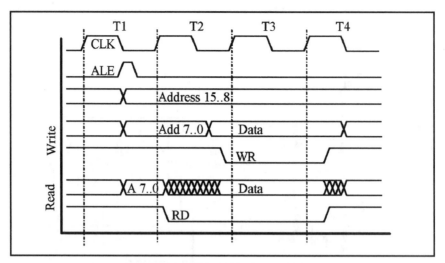

Figure 10-6. Memory access timing.

When the Address-Latch Enable (ALE) line goes high, the 74HC573 latch becomes transparent to its inputs. Any data presented to the inputs is passed directly to the outputs. ALE soon goes low again, latching the address information from Port A. The full address is now available to any device, with the high byte in the latch and the low byte at Port C. Though it is not necessary for smaller circuits, a second latch can be placed on Port C to buffer and amplify the address. The 74HC573 presents a cleaner and stronger signal than the MCU, allowing it to drive more devices.

In a write cycle, Port A switches from the high address byte to the data byte, and the Write Enable pin goes low, signaling any peripherals on the bus to accept the data.

In a read cycle, Port A switches from address out to data input mode. The Read Enable pin goes low, signaling any peripherals on the bus to send their data.

The only device in this simplified schematic is the Hitachi-style 62256 32K by 8-bit memory chip. In this configuration, the 62256 holds memory from address 0x0000 to 0x7fff, though the MCU doesn't use it for memory locations 0x0000 to 0x0260. A useful circuit-board layout trick: the address output lines A14 through A0 from the MCU do not necessarily need to connect to the address input lines on the memory in that same order. As long as reading and writing are consistent, it doesn't matter if address 0x0123 in the MCU is actually address 0x0312 in the memory.

There are several different memory devices that can be plugged into this schematic. For example, the Hitachi HM62256B, or the fully compatible Mosel-Vitelic V62C518256. Both of these (or any other compatible 32K static RAM) provide high-speed, though volatile, storage for half of the Atmel's address space. Unfortunately, the use of static

RAM makes the robot less than portable. Once power is removed from the circuitry, the program is lost and must be reloaded from the desktop computer.

Dallas Semiconductor has a pin-compatible nonvolatile static RAM, the DS1230Y/AB, which adds a small lithium battery to the memory package. This battery can keep the memory data intact for up to ten years without external power. Another nonvolatile solution is a new technology from Ramtron, their FRAM memory. Unfortunately, writing to FRAM is significantly slower than writing to the other static RAMs.

The use of external RAM removes 18 bits of I/O from the MCU. This is a painful loss for a robotics system which needs to sense and control a rich environment. However, the external RAM architecture also opens up a new method of hardware interfacing known as memory-mapped I/O. Just as most devices can be found in SPI or serial interface form, most devices can also be found with MCU addressable interfaces. Since the 62256 memory chip only uses the bottom 32K of address space, that leaves the top 32K of address space for use by peripherals. A complete MCU system is shown in *Figure 10-7.*

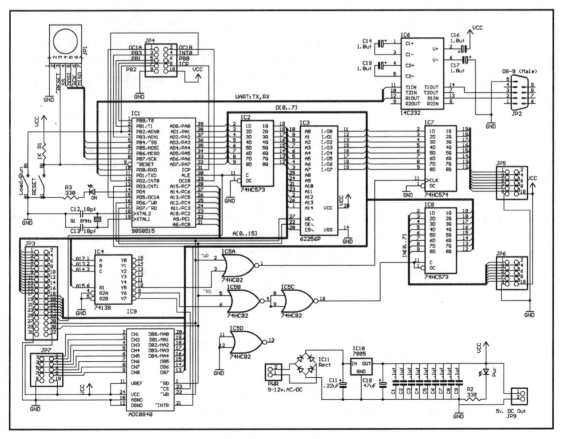

Figure 10-7. Complete Fuzbol system.

This system is complex and intimidating at first glance. It consists of six simple sections, each of which uses technology we have already seen. The schematic and PC board layouts are available on the SRS website, and these boards are also available from SRS.

MCU and Memory

The first section consists of IC1, IC2, and IC3. This is nothing more than the MCU with external RAM that was covered previously. As a space saving convention, sets of related signals (such as the SPI lines) are grouped into a single thick line representing a data bus.

Note that a variety of pins from the MCU are exported to a 2 by 5-pin header, JP4. The PWM outputs, an unused interrupt, and some Port B pins. This allows you to use these lines in an external circuit. JP1 is of course the SPI interface, the now familiar DIN-7.

Address Decoder

The second section of the circuit is the address decoder (IC4 and IC5), a required aspect of any memory-mapped system. The 74HC138 decoder/demultiplexer serves two purposes. First, it has a complex system of enable and latch signals (G1, G2A, and G2B). Second, it converts three bits of address (A, B, C) into a low signal on one of its eight outputs (Y0 through Y7).

When the latch enable G1 is high, the decoder is active and any changes in the address inputs affect its internal state (and outputs, if the output is enabled). Once G1 transitions from High to Low, the address is locked. Address bit 15 controls G1. When it is low, the SRAM is active. When it is high, the peripherals are active.

G2A and G2B are active-low output enables. When G2A or G2B are high, all outputs are disabled (set high, since the 74HC187 is an active-low device). G2A and G2B are both hard-wired to ground, allowing all enable control to be driven by address bit-15 on G1. Once the address reaches 0x8000 or above, the decoder output is enabled and devices tied to it downstream may use the address and data bus.

G2A and G2B are wired to ground in this circuit, but for a more complex system they can be used to cascade this device from a higher-level 74HC138 or off of a 74HC688, which is useful for matching eight bits of address at once.

IC5, a simple quad NOR gate, is used to combine read and write signals with the decoder enable outputs. If either the decoder or the memory access signal is high, the output of the NOR is low. Another NOR in series inverts this signal back to an active low.

Communications

IC6 is the RS-232 level converter and its signals are communicated through DB-9 connector JP2.

Digital I/O

Having taken away most of the I/O ports with memory interfaces and serial communications, it is necessary to give them back. IC7 and IC8 work in conjunction with IC5 to do just that.

IC7 is a digital output only. The 74HC574 tri-state flip-flop latches data on a rising clock. This is provided by IC5-A when the address is 0xDxxx (anywhere form 0xD000 to 0xDFFF) and the write enable line goes low. Output enable is wired directly to ground, so the port always presents valid data to its jumper JP5.

IC8 is a digital input only. The 74HC573 is transparent whenever its latch enable is high, and locks that data into place when the latch is low. The output is only enabled when the latch control pin transitions to low (it is edge-triggered). With the latch and output both wired together, whenever the control signal is high any data coming in from JP6 may change, but it is not visible to the bus. When the control signal goes low, the data is fixed and visible. IC5-B and IC5-C combine the encoder's signal with the read enable so that whenever the address is 0xExxx and read enable is low, the control signal is low.

Although there are chips that allow both input and output on the same address (such as the 74HC245, 74HC646, or 74HC652), it is simpler to work in only one direction at a time.

Analog Input

IC9 is the MCU compatible version of the serial A/D converted used in previous projects. The ADC0848 is enabled whenever the address is at 0xFxxx, so that any read or write from memory address 0xF000 to 0xFFFF will activate the converter. This chip is well suited for the Atmel MCU. It has active-low read and write signals, allowing it to interface directly with the MCU.

The A/D converter is wired into one of the two external interrupts on the 8515. This line is pulled low whenever a conversion is completed, signaling the MCU. This leaves only one other interrupt for external devices, such as the feeler switches. Since there are two feelers, they need to be OR'ed together so that if either one of them is touched, an interrupt is generated. In addition to this, the feeler outputs can be sent directly to one of the input ports so that the interrupt handler can determine which of the feelers has triggered the interrupt, as shown in *Figure 10-8*. This technique can be extended to combine any number of interrupt sources.

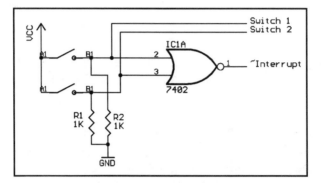

Figure 10-8. Feeler interrupts.

Additional I/O

The remaining I/O ports on the MCU are exported to JP4. This includes the PWM ports OC1A (which is also Port D Pin 5) and OC1B, and Port B pins 0, 1, 2, and 3. These ports are used in later projects.

The entire address and data busses, plus the memory control signals, are exported on JP3. This allows any number of additional memory-mapped devices to be controlled from this board.

Power Supply

The sixth and final section is the power supply. This supply is rudimentary and can handle about one amp of power. It demonstrates a few nice touches for any power supply.

The first thing to build into any supply is a rectifier (IC11) on the input. Though the diodes can cause up to 1.5-volt dropout, they also protect the circuit from "bad" power inputs. I used an MCU board once that did not have a rectifier on the input, and when I switched to a different wall transformer I ended up destroying the MCU. The polarity was opposite in the new transformer's power plug.

IC10 is the standard voltage regulator, and can be replaced with the voltage management device of your choice.

Finally, the string of capacitors C1 through C9 are not actually clustered on the power supply. They are spread across the entire circuit and act as filters for the digital chips.

Every chip on the board has its own personal .1uf capacitor to smooth over power glitches introduced during the chip's operation.

Program 10-5: Fuzzy Braitenberg

With this new, expanded, architecture it is now possible to explore a wider array of robot behaviors. The basic Braitenburg behaviors from Chapter 6 are reproduced here in Fuzbol. The full program listings can be found on the CD.

Supporting Code Modules

With external memory, there is less need to make code as small and efficient as possible. With the new architecture, it is now possible to make complex programs. With this new complexity, it becomes desirable to build these programs from several small (and simple) modules. There are two Fuzbol modules that the fuzzy Braitenberg programs all have in common, Motor.fuz and ioPort.fuz.

Motor.fuz

The motor controller object manages the PWM system in the MCU and the Reverse lines. It assigns the PWM output OCR1A to the left motor, and OCR1B to the right motor. Of course, PWM uses Timer 1 so no other Timer 1 modules can be active when the motor is running. Port B Pin 0 is assigned to Left Reverse, and Port B Pin 1 is Right Reverse.

Motor:Init() starts the PWM system with a speed of 0 for both motors, and both reverse lines are set low (forward).

Motor:LeftRev() and Motor:RightRev() each take one Boolean value (0 or 1), and set the left or right reverse lines appropriately. Motor:Stop() sets both PWM lines to zero, and both direction lines to forward.

ioPort.fuz

The I/O Port object manages the three memory-mapped devices on the extended MCU board.

ioPort:DigitalOut maps to the address of the output latch at address 0xD000, and is a write-only port. ioPort:DigitalIn maps to the input latch at 0xE000, and is a read-only port.

ioPort:Analog[] is a data array, storing the last known conversion values for each analog input channel. This array is accessed directly, indexed to the channel of interest.

Supporting the analog conversion process is the address ioPort:a_to_d, which is used to communicate with the ADC0848 chip. This should not be used by applications using this module, since the ioPort:Convert() handler is constantly reading or writing ioPort:a_to_d.

ioPort:Convert() operates in a loop, converting each channel of analog input in order and storing the results in the ioPort:Analog[] array. Each channel takes about 40 uS to convert, so all eight channels are converted once every 320 uS or so. This is about 3,000 conversions per second per channel which is more than enough for any application this system will be used for.

ioPort:Init() sets up the interrupt and ports for the module, and begins the conversion process.

Each of these vehicles uses the same external sensors as the previous Braitenberg vehicles. Two photoresistors which are attached to channel 1 and channel 2 of the analog input port, and two feeler switches which will be attached to the digital input port pins 0 and 1.

Vehicle 1

Timid

The basic framework for Timid, and all of the other behaviors, is the Heartbeat.fuz program. The start() function in Timid.fuz initializes the heartbeat, ioPort, and Motor drivers. Timid:Init() starts the heartbeat interrupt, and Timid:Pulse() contains the robot's behavior code and is executed once per pulse.

```
Timid:Pulse()
{
        left_eye as byte;
        right_eye as byte;

        beat = beat - 1;
        if (beat <= 0)
        {
           beat = PULSE_CNT;
           System:PORTD = System:PORTD ^ ON;

           left_eye = ioPort:Analog[0];
           right_eye = ioPort:Analog[1];

           if (bright == 0)
         {
                bright = (left_eye max right_eye) + 4;
        }
        else
        {
                if ( (left_eye > bright)
```

```
                           | (right_eye > bright) )
                       {
                           Motor:LeftVel = FULL_SPEED;
                           Motor:RightVel = FULL_SPEED;
                       }
                       else
                       {
                               Motor:LeftVel = FULL_STOP;
                               Motor:RightVel = FULL_STOP;
                       }
                   }
               }
}
```

The local variables left_eye and right_eye are not strictly necessary, but they provide an appropriately named storage area for the analogous Analog[] array entries. Though this is inefficient, it is a minor inefficiency and adds to the readability of the code. The first pass at writing any program should aim at making clear, easily read code. If that code then proves to be too slow or too large, then a second pass can be made to trim it back or to re-think the algorithms for speed instead of readability.

When the program is first run, the variable bright is set to zero. This allows the test "if (bright == 0)" in Pulse() to evaluate to TRUE on the first heartbeat. Bright is then assigned the maximum light level detected by the two eyes, plus a small margin. Any subsequent light level that goes above this preset is then considered bright.

If either eye sees bright light, both motors are activated. When neither eye is in the sunshine, the motors are turned off.

Since this code is self-calibrating, all you need to do to make it work is place the robot in the shade and start the program. It will happily sit there in the dark until light intrudes, and then it will run blindly until it finds the dark again.

Indecisive

To turn Timid into Indecisive, the central control code is adjusted so the robot is always moving. The bright/dark decision controls the direction instead.

```
if (bright == 0)
{
        bright = (left_eye max right_eye) + 4;
                Motor:LeftVel = FULL_SPEED;
                Motor:RightVel = FULL_SPEED;
}
else
{
        if ( (left_eye > bright)
        | (right_eye > bright) )
        {
        Motor:LeftRev( 1 );
        Motor:RighRev( 1 )
```

```
            }
            else
            {
            Motor:LeftRev( 0 );
            Motor:RighRev( 0 )
            }
        }
```

This program will switch the relays a lot, so you may not want to run it for long.

Vehicle 2

These vehicles use two eyes, and control the motors independently.

Fear

Born in the dark, this creature is afraid of light and will run away from it.

```
        range BLACK      ( 0x00, 0x00, 0x10 );
        range DARK       ( 0x20, 0x00, 0x18 );
        range DIM        ( 0x40, 0x00, 0x20 );
        range BRIGHT     ( 0x90, 0x00, 0x50 );
        range LIGHT      ( 0xF0, 0x10, 0x50 );

      rule MoveLight
    {
      BLACK    as 0x00;
      DARK     as 0xA0;
      DIM      as 0xC0;
      BRIGHT   as 0xF0;
      LIGHT    as 0xFF;
    }
```

Fear.fuz uses the Fuzbol Range and Rule instructions. The brain is self-calibrating at startup, like Timid was. The ranges are balanced to have more control at low light values. These ranges determine how sensitive the fuzzy rule is across the input scale. Where detailed control is desired, there should be more ranges with narrower extents.

The rule MoveLight then uses the ranges and specifies the velocity curve across the span of inputs. This rule sits still for only a very narrow span of true darkness, and accelerates to near-full velocity quickly from there, as shown in *Figure 10-9*.

By changing both ranges and rules, the shape of this velocity curve can be altered to suit any whim. It is important to remember, however, that the light values will not reach the maximum of 255, since they are offset down by the dark value. The sensors will not return a full-range of values under most circumstances anyway, so most of the control occurs in the lower half of the rule set.

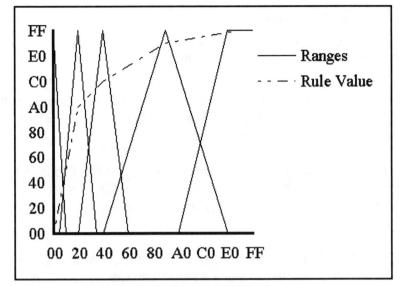

Figure 10-9. Ranges and velocity curve.

```
if (dark == 0)
{
        //
        // At first time through, decide what "dark" is
        //
        dark = (left_eye min right_eye);
        if (dark > 4)
        {
                dark = dark - 4;
        }
}
else
{
        //
        // Calibrate the current eye values relative to
        //      relative "darkness"
        //
        if (left_eye < dark)
        {
                left_eye = 0;
        }
        else
        {
                left_eye = left_eye - dark;
        }
        if (right_eye < dark)
        {
                right_eye = 0;
        }
        else
        {
                right_eye = right_eye - dark;
        }
        //
        // Run away from the light
```

```
//  (actuate the motor near the light proportional
//       to the light level)
//
Motor:LeftVel = left_eye is MoveLight;
Motor:RightVel = right_eye is MoveLight;
}
```

The first pass through the heartbeat reads the two eyes. The darkest of these eyes (less a small margin) is then set as the preferred light level. Anything this dark or darker will read as no light to the robot.

Every other pass through the Pulse() handler, each eye's reading is reduced by this base-level darkness. Then the actual work is done by the rule MoveLight. The left motor is activated according to the left eye's reading, and the right motor by the right eye.

To test this behavior, set the robot in the middle of a room and start the program. It helps if both eyes are receiving roughly the same light when it starts, otherwise it may start spinning until both eyes are in the dark. Now shine a flashlight on the eyes and see how it reacts.

The self-calibration can be tested by stopping the program, carrying the robot to a room with a different light level, and restarting it. It should behave essentially the same, compensating for the ambient light levels.

Aggression

Aggression is not a simple reversal of the Fear motor/eye pairing (though that is the basis of the change). This program actively seeks out light with the intent of smashing it.

The ranges and rules have been changed, so the robot never really stops but is always crawling forward looking for light. This could be a problem, since the feelers are not hooked into any logic yet.

Once light is detected, the robot accelerates towards it until it goes way. If the light merely moves, the robot will turn to follow it.

Load the program Aggression.fuz and see how its behavior is different from Fear.

Vehicle 3

Love and Explore

Love.fuz has the same control code as Fear.fuz, but the motor assignment is re-versed. The robot races at high speed in the dark and slows as it approaches light. When it is bright enough (which may take a very bright light), it stops.

```
Motor:LeftVel = 255 - (left_eye is MoveLight);
Motor:RightVel = 255 - (right_eye is MoveLight);
```

This program needs more tuning than previous ones. Where it can self-calibrate dark for the ambient light level, the maximum illumination it gets will be different for different lights.

```
dark = dark min (left_eye min right_eye);
```

Another improvement is a dynamic sense of darkness. Dark constantly tracks the lowest level of either eye, which simplifies the code by removing the test for "dark = 0."

It is also possible to track the highest value of light, and then scale the photocells to a full range using the dark and light values, except that requires more math. Math in Fuzbol is very slow; division the worst, then multiplication. The slowest of all is real-valued math and it is best to avoid it unless necessary.

The Explore behavior would simply reverse the left and right motor hookups from Love.fuz.

Dogged

Dogged.fuz takes Love.fuz and adds additional feeler code. The robot drives ahead, faster in the dark and slower in the light. If it hits something it backs up for a while before returning to the light seeking behavior.
For this behavior, the feeler switches must be attached to DigitalIn pins 0 and 1. The two feelers are also mixed with a NOR gate, and the result is sent to Interrupt-0 so the robot will respond immediately to any contact.

There are a number of enhancements in this program.

```
if (count_rev == 0)
{
        //
        // Only seek light when not escaping obstacles
        //
        Seek( left_eye, right_eye );
```

```
}
Reverse( ioPort:DigitalIn & 0x03 );
Update();
```

The brainwork in Pulse() has been broken out into three separate functions. Seek() sets target speeds (left_go and right_go) for the motors based on the incoming light. Seek() is only called when the reverse counter is zero, though it could be called every pulse with the same effect.

After Seek(), Reverse() checks the state of the contact feelers. If one or both of the feelers have been touched, it reverses both motors and sets target velocities that will cause the robot to back up in an arc away from the obstacle. It also sets the reverse counter. This counter forces Seek() to be disabled for a specific number of heartbeats.

Update() takes the target velocities and applies them to the motors.

```
Motor:LeftVel = (Motor:LeftVel + left_go) >> 1;
Motor:RightVel = (Motor:RightVel + right_go) >> 1;
```

The improvement in Update() is that it implements an acceleration curve, stepping halfway between the actual velocity and the target velocity each pass. The drawback to this particular form of acceleration curve control is that it is using byte values. If the sum of the motor velocity plus the "go" amount is ever greater than 255, the addition will wrap-around to zero causing irregular speeds. If you find your robot starting and stopping when it should be smoothly accelerating, check any math involving byte variables.

```
Robot:Bump()
{
        //
        // On bump, come to a dead stop
        //
        if (count_rev == 0)
        {
                Motor:LeftVel = 0;
                Motor:RightVel = 0;

                left_go = 0;
                right_go = 0;
        }
}
```

Bump() is the handler for interrupt 0. When the robot bumps into anything, it forces the motors (and target velocities) to zero, but only if the robot is not already backing away from an obstacle.

Noise from the motors can be picked up all over this circuit. It can be picked up on the photoresistors, even on the feeler switches. False signals can cause all types of problems in a robot, from false readings on feelers, to resetting the MCU. Careful and liberal use of capacitors can help reduce the problem.

Insecure

The insecure robot starts out as a light tracker, looking for the brightest light and moving towards it. It is aggressive and easy to control with a flashlight. However, if this robot goes without touching an obstacle for too long it grows insecure. It begins to ignore the light, and instead travels in a tighter and tighter curve to it's right. Finally, if no wall is found it panics and runs off straight ahead in the hope of finding the security of a wall or other obstacle. Insecure.fuz implements these behaviors and uses fuzzy variables for the first time. This program also introduces a new concept to the robot, an internal sense of well-being. There can be any number of internal senses to drive the robot to fulfill whatever needs you design into it.

```
range BLACK     ( 0x00, 0x00, 0x10 );
range DARK      ( 0x20, 0x00, 0x18 );
range DIM       ( 0x40, 0x00, 0x20 );
range BRIGHT    ( 0x90, 0x00, 0x50 );
range LIGHT     ( 0xF0, 0x10, 0x50 );

rule MoveLight
{
  BLACK    as 0x20;
  DARK     as 0x90;
  DIM      as 0xC0;
  BRIGHT   as 0xE0;
  LIGHT    as 0xFF;
}

range HAPPY     ( 0x00, 0x20, 0xD0 );
range LOST      ( 0xA0, 0x00, 0x60 );
range PANIC     ( 0xF0, 0x10, 0x70 );
```

Insecure uses essentially the same MoveLight rule, with essentially the same range definitions as the previous behaviors. The MoveLight motor driver curve starts at 0x20, however, and has a very steep slope. Three additional ranges have been added, one for each mood range.

```
Robot:Pulse()
{
        left_eye as byte;
        right_eye as byte;

        beat = beat - 1;
        if (beat <= 0)
        {
                beat = PULSE_CNT;
```

```
        System:PORTD = System:PORTD ^ ON;

        left_eye = ioPort:Analog[0];
        right_eye = ioPort:Analog[1];

        //
        // Continuously update dark
        //
        dark = dark min (left_eye min right_eye);

        //
        // Calibrate the current eye values relative to
        //      relative "darkness"
        //
        left_eye = left_eye - dark;
        right_eye = right_eye - dark;

        //
        // Manage the security blanket
        //
        if (security < 255)
        {
                security = security + 1;
        }

        //
        // Behaviors!
        //
        Reverse( ioPort:DigitalIn & 0x03 );
        if (count_rev == 0)
        {
                clear( left_go );
                clear( right_go );

                left_rev = 0;
                right_rev = 0;

                SeekLight( left_eye, right_eye );
                SeekWall();
                Panic();
        }
        Update();
    }
}
```

Init() sets up the various variables and interrupts. Pulse() performs all the action, timed off of the heartbeat timer interrupt. The left_eye and right_eye variables get their values from the analog-to-digital converter, and are referenced to the darkest light the robot has ever seen. A new variable, security, is added. Security increments each pulse, up to its maximum value, and moderates each of the seeking behaviors.

Reverse() is called each pulse to check the feelers. If something has been touched, the Reverse() function resets security to zero and sets up the backup behavior.

When the robot is not backing up, each of the searching behaviors are evaluated. The Robot variables left_go and right_go are defined as fuzzy instead of byte or integer. Before left_go and right_go can be adjusted, they must be cleared. It is the nature of fuzzy variables that assignments to those variables do not completely reset their contents. Fuzzy assignments accumulate the values into the variable to the extent of the current system weight. If the system weight is high, the assignment effects the fuzzy variable a lot. If the system weight is low, it will not affect the variable much at all.

SeekLight(), SeekWall(), and Panic() each assign values to the left_go and right_go fuzzy variables in their own way. Update() defuzzifies the go variables and sets the motor velocities accordingly.

```
Robot:SeekLight(
        left_eye as byte,
        right_eye as byte )
{
        weight( security is HAPPY );
        //
        // Wander until the light is found
        //
        left_go = right_eye is MoveLight;
        right_go = left_eye is MoveLight;
}
```

SeekLight() is simple. To the extent that the security counter is in the range HAPPY, it will drive towards bright lights. Since weight() is set inside the SeekLight() function, it only affects assignments within that function.

```
Robot:SeekWall()
{
        weight( security is LOST );

        //
        // When feeling lost, curve to the right...
        //
        left_go = 0xF0;
        right_go = 0x40;
}
```

To the extent that the security counter is in the LOST range, SeekWall() will drive the robot in right-turning circles.

```
Robot:Panic()
{
        weight( security is PANIC );

        //
        // When feeling REALLY lost, zip straight ahead!
        //
        left_go = 0xFF;
        right_go = 0xFF;
}
```

Panic()drives the robot straight ahead at full speed, but only to the extent that the security counter is in the PANIC range.

Insecure demonstrates two forms of behavior layering. When Reverse() sets the count_rev variable, all of the seeking and panicking behaviors are disabled for the duration of the reversal. Reverse() is an overriding behavior, on the level of a life saving reflex. The seek and panic behaviors, when active, blend gently into each other depending on the context of the robot. In this example, the context is defined by the amount of time since the feelers touched something.

The use of the fuzzy rule MoveLight, plus the fuzzy variables left_go and right_go, give the robot complex behaviors with only a few lines of code.

Vehicle 4

Most of the behaviors in the previous vehicles were based on simple likes, simple dislikes. You can also create behaviors based on non-linear responses to input. There can be complex interlocking drives and impulses, each with a preferred area of activation.

Moth

A moth robot will be attracted to moderate light, but will be repulsed by a very bright light. The heart of this behavior lies in the MoveLight rule. At the extremes of darkness and bright light, the robot is inhibited. Only in the twilight shadows does it truly come alive.

```
        range BLACK      ( 0x00, 0x00, 0x40 );
        range DARK       ( 0x40, 0x00, 0x40 );
        range DIM        ( 0x80, 0x00, 0x40 );
        range BRIGHT     ( 0xC0, 0x00, 0x40 );
        range LIGHT      ( 0xFF, 0x00, 0x40 );

    rule MoveLight
    {
      BLACK    as 0x40;
      DARK     as 0x98;
      DIM      as 0xC0;
      BRIGHT   as 0xFF;
      LIGHT    as 0x00;
    }
```

The ranges are uniform and unexciting, and the rule carries all of the behavior. Non-linear preferences can be fancier than this. By adding additional detailed ranges, the rules can describe complex preference functions.

```
Robot:Pulse()
{
        left_eye as byte;
        right_eye as byte;
        scale as real;
```

```
        beat = beat - 1;
        if (beat <= 0)
        {

                beat = PULSE_CNT;
                System:PORTD = System:PORTD ^ ON;

                left_eye = ioPort:Analog[0];
                right_eye = ioPort:Analog[1];

                //
                // Continuously update dark
                //
                dark = dark min (left_eye min right_eye);
                light = light max (left_eye max right_eye);
                scale = 255.0 / (light-dark);
                if (scale < 1.0)
                {
                        scale = 1.0;
                }

                //
                // Calibrate the current eye values relative to
                //       relative "darkness"
                //
                left_eye = (left_eye - dark) * scale;
                right_eye = (right_eye - dark) * scale;
                //
                // Behaviors!
                //
                Reverse( ioPort:DigitalIn & 0x03 );
                if (count_rev == 0)
                {
                        left_rev = 0;
                        right_rev = 0;
                        SeekLight( left_eye, right_eye );
                }
                Update();
        }
}
```

The main Pulse() function is essentially the same as in the other light-seeking robots. The main enhancement is the way it scales the input light values. Earlier robots simply applied a dark offset. This robot tracks both the minimum and maximum light levels (dark and light), and then scales the current light values at each eye. The offset and scaling process normalizes the value so it covers the full scale of 0 to 255. Any light level that is at dark is transformed to 0x00, and any light level that is at light is transformed to 0xff. Other light levels are scaled to appropriate intermediate values.

This uses real-valued math, which is very slow. In this case, the sensor scaling is central to the robot's behavior. The light values it operates on must cover the entire range.

```
Robot:Update()
{
        left_vel = (left_vel + left_go) >> 1;
```

```
        right_vel = (right_vel + right_go) >> 1;

        Motor:LeftVel = left_vel;
        Motor:RightVel = right_vel;

        Motor:LeftRev( left_rev );
        Motor:RightRev( right_rev );
}
```

The update function has been improved again. The variables left_vel and right_vel are used to handle the acceleration curve as usual. Since they are integers and not bytes, there is no chance of them overflowing at large velocities, an improvement from earlier versions.

Vehicle 5 and more

There are 14 vehicle types described in Braitenberg's book. Vehicle 5 discusses the use of logic to control the robot. Where earlier vehicles use threshold logic to excite or inhibit motor action, a more complex controller can mix, sum, and sequence multiple threshold systems to create intricate behaviors. State machines and simple neural networks demonstrate this behavior.

Other vehicles introduce the concepts of natural selection, evolution, and the use of high-level concepts to represent things in the world. Some of the concepts touched on include motion detectors, eyes and cyclic thinking, though few of them lend themselves to easy implementation.

Following

The robots' behaviors are all designed to follow or avoid light. This gives you the builder, a form of photonic leash guiding your creation with a flashlight or with shadows. The same ideas, mechanisms, and code can be adapted to follow other things as well.

Try creating a robot that follows a wall using either a photocell pointed towards the wall, or a switch feeler that the robot keeps in contact with the wall.

A preference to turn towards darkness can be converted to the ability to follow a black line on the floor by pointing the eyes down. This may require special illumination from the robot itself, but the idea of tracking a black line is not much different than tracking a flashlight.

By building two or more robots, you can initiate a variety of robot games. Place a modulated light on each robot. Mount some IR detector modules at the head ends of the robots, and the robots can see and track each other. What games can robots play? Tag? Hide-and-seek? Soccer?

CHAPTER 11
More Sense

Where am I? Position Mapping

One of the first questions an intelligent computer program will ask of its mobile hardware is, "Where are you?" This is actually a very difficult question to answer. Even if the robot has a GPS (Global Positioning Satellite) unit attached, it would only know its position to the nearest 10 yards or so.

For a simple behavior-oriented robot, its location in space doesn't matter. The purpose of the reactive robot is not to get to point B from point A, but to respond to its environment in a reasonable manner. As the level of abstraction and the need for control rises above the level of pointless wandering, it becomes necessary to know where the robot is, where it has been, and, hopefully, where it is going.

There are a number of ways for a robot to know exactly where it is, other than GPS. The room (or rooms) the robot lives in could have beacons installed in the corners of the ceiling. Sensors on the robot can triangulate its position from these beacons *(Figure 11-1)*.

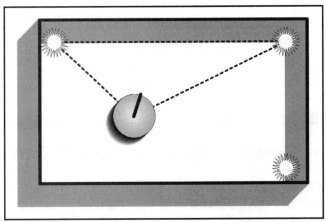

Figure 11-1. Triangulation.

Doorways could have bar codes marked on the wall or floor, so the robot only has to traverse the wall until it can read its location from the codes *(Figure 11-2)*.

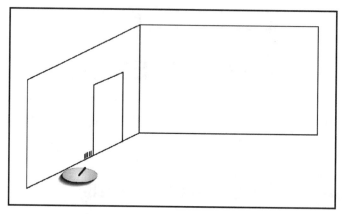

Figure 11-2. Marked environment.

Without marking the environment, the robot could detect certain features (such as doorways, corners, and steps) of its environment using sensors and try to correlate them with the memorized or pre-programmed layout of known rooms *(Figure 11-3)*.

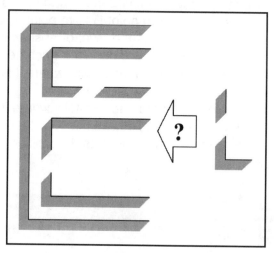

Figure 11-3. Matching map sections.

The next, and fortunately easier task is for the robot to know where it has been.

Where Have I Been? Leaving a Trail of Crumbs

Even when a robot starts from a known position or doesn't care where it starts because it is mapping the environment in blank memory, it might still need to know where it has been. Note that no amount of robot tracking and mapping eliminates the need for sensors and some form of reactive behavior to keep the robot out of trouble. All forms of position sensing and tracking have errors, though various researchers have made careers out of compensating for them. Most environments are also full of hazards leaping into the path of the robot.

People with lots of money outfit their robots with inertial sensors and gyroscopes so that every motion and wiggle that occurs to the machine is recorded and accounted for. These included such robotic machines as the auto-pilot in better airplanes and spacecraft. Researchers also place Doppler radar on the corners of the robot to measure how fast the ground is moving underneath it.

The rest of us use a form of dead reckoning known as odometry. In odometry, you measure the motion of the wheels and calculate the robot's change in position. There are several types of sensors available to sense rotary motion, from analog resolvers (which measure the angle of a magnetic field), to absolute optical encoders (which output a binary number indicating the current angle of the object in question), to incremental optical encoders (which only report the relative angular motion).

Odometry, the art of measuring distance

First, a caveat. Measuring the motion of a robot's wheels is not the same as measuring the motion of the robot. If the contact point of a wheel is at all wide, it will slip unpredictably when the robot turns. If the robot accelerates or decelerates quickly (or drives across uneven terrain, or simply moves at all) there will be some slipping between the wheels and the ground. Finally, the granularity of the measurements affect the accuracy with which you can track the robot's motion. All tracking errors are compounded when the robot turns. After a turn, all forward motion magnifies any errors suffered measuring the angle of the turn.

Errors can be reduced by using thin wheels to reduce the contact area to a mathematical point. You can also measure motion with extra spring-loaded wheels which are unlikely to slip (instead of the drive wheels), and use high-resolution encoders. But still, it won't be perfect.

Second, another caveat. The system of odometry described here works for differential drive robots. It assumes that there are two central drive wheels which drive and turn the robot, and a third (or possibly more) castors which serve only to balance the robot, Tank-steering. For car-steering robots and other, more complex, drive systems this math will not apply so gracefully.

From these auspicious beginnings, we launch into an example of building odometry into your robot. This is not an actual project chapter, but an introduction to the issues involved.

Incremental Optical Encoders

Incremental optical encoders are the least expensive encoder to buy and the easiest to build from spare parts. These encoders consist of a specially marked disk like the one shown in *Figure 11-4*, and an optical sensor. The encoder disk shown here has 64 segments (32 white and 32 black), which gives 32 high pulses and 64 level transitions. Commercial encoders can have from 16 to 1024 or more segments. The more segments, the finer the measurement possible.

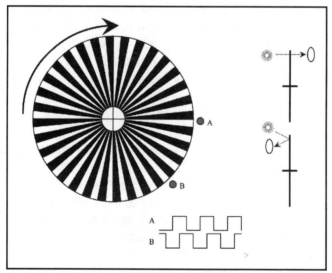

Figure 11-4. Optical encoder.

There are two ways to detect the light and dark patterns on the encoder disk: transmission (top diagram) and reflection (bottom diagram). In transmission encoders, the black stripes are printed on a transparent material. An optical gap detector senses the clear areas as signal and the black areas as no signal. Optical gap detectors have an LED on one side, and a photodetector on the other side. They are available in many sizes and configurations. For reflection encoders, the LED and detector are on the same side of the disk. The white areas reflect light back to the detector and the black segments absorb the light.

For either type of detector, you can print the pattern on clear plastic or heavy white paper with a laser printer. Once the pattern is printed, protect it with clear tape or lamination so it doesn't scratch off through accidental contact with the detector or environment.

The encoder can be coupled to the drive (or measurement) wheels through any number of mechanisms. The simplest way is to buy a motor with the encoder built into it. If you are adding an encoder to an existing system, try to place it on the motor shaft before it enters the gear train to get the maximum number of pulses per drive-wheel revolution. At the other end of mounting complexity, you can attach a reflective encoder disk to the wheel itself (or paint it onto the wheel's hub). With this form you can measure the wheel's motion directly.

An optical encoder with one detector can measure the motion of the disk, but it has no way to detect which direction it is moving. In a mobile robot the direction of the motors is under fairly direct control of the MCU, so this is not a critical failing. Unfortunately, in a single detector system, when the disk is moving very slowly (or the robot is stopped), it is possible to get false triggers from the detector. As the disk slowly drifts past the detector, the boundary between light and dark reaches a point where it is just at the threshold of activation. Any vibration or electrical noise can induce a rapid stream of signals.

Most commercial encoders use two detectors positioned like (A) and (B). In the configuration shown in *Figure 11-4* these give pulse trains that are 90 degrees out of phase, known as quadrature encoding. This doubles the number of level transitions per revolution (A plus B transitions), and can give a finer measure of the disk's velocity. The system can also tell which direction the wheel is moving. If (B) is low and (A) transitions from low to high, it is moving one direction. If (B) is low and (A) goes from high to low, it is moving the other direction. If (A) is oscillating on a segment boundary and generating a pulse stream, then the position counter simply moves up and down one bit.

There are special purpose encoder chips that watch the quadrature signal and provide useful position and velocity information. Though it doesn't make full use of information from the encoder, you can use a simple edge-triggered binary counter such as the 74F269, 74F579, or 74F779 with the (B) signal wired to the Up/Down direction pin, and (A) wired to the clock pin. The MCU can then read the 8-bit position value as desired. It is also possible to avoid extra hardware by attaching one or both of the encoder signals to an interrupt on the MCU, and count the pulses in software.

Pulse to Distance

Once you have attached the encoder hardware to the drive system, and have a method of tracking the pulses from the encoder, it is necessary to convert that pulse train into a velocity or a measure of actual distance moved. For this phase, we assume the system is counting both high-low and low-high transitions on only one sensor. We ignore the subtleties of direction sensing and error rejection.

First, determine how many pulses the MCU will count for each full rotation of the drive wheel. If the encoder is attached to the drive axle or wheel itself, this is the pulse-per-revolution count of the encoder itself. If the encoder is upstream from a gear-train, then the pulse-per-revolution of the encoder must be multiplied by the gear reduction to get the pulse-per-revolution of the drive wheel.

Second, calculate the distance traveled in a straight line during one full revolution of the wheel. This is simply the circumference of the wheel, π times the diameter of the wheel.

For example, for a 64-pulse encoder attached directly to a 2 ½" diameter wheel:

$$\pi \times 2.5 = 7.85 \text{ inches per revolution}$$
$$7.85 / 64 = .12 \text{ inches per pulse, or roughly 1/8" per pulse}$$

If the brains count 100 pulses from a wheel, that wheel will have moved 12.27 inches (give or take slippage).

It is also possible to use an internal timer to measure the duration between pulses to give the velocity of the wheel. This velocity measure could then be used as feedback to lock the wheel into a specific rate of rotation. Note that this velocity calculation works best for a rapid pulse stream. On a low frequency signal the timer can easily overflow.

Wheel Motion to Robot Position

Given the distance traveled by each wheel over a given interval of time, it is possible to calculate the change in position of the robot's center for that same time interval *(Figure 11-5)*. The center of the robot is that point which lies directly between the contact point of each wheel. On rough terrain, the contact point shifts as the robot travels over bumps so the center also changes, adding error to the calculation.

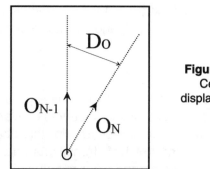

Figure 11-5.
Center
displacement.

Given the signed distance (positive forward, negative backwards) traveled by the left wheel D_L and the right wheel D_R, the linear distance traveled by the robot's center D_C is calculated by:

$$D_C = (D_R + D_L) / 2$$

Given the width W between the contact points of the robot's wheels, the change in orientation D_O *(Figure 11-6)* in radians is calculated by:

$$D_O = (D_R - D_L) / W$$

The robot's new orientation O_N is updated by D_O:

$$O_N = O_{N-1} + D_O$$

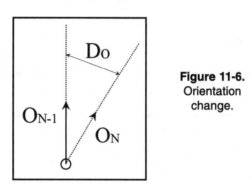

Figure 11-6.
Orientation
change.

The robot's new center position X_N and Y_N in space *(Figure 11-7)* is calculated by:

$$X_N = X_{N-1} + D_O * \cos(O_N)$$

$$Y_N = Y_{N-1} + D_O * \sin(O_N)$$

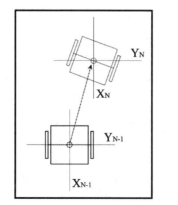

Figure 11-7.
Center
coordinates.

When the wheels are turning at different rates, the robot is traveling a curved path. The math is calculating the position based on linear movement across the chord of this arc, introducing more errors to the position. This error can be reduced by using a very fine encoder and updating the position at frequent intervals.

What has happened to me? Event Mapping

Now that it is possible to trace the relative position and orientation of the robot through space, what good is this information? On a simple aesthetic level, the robot could radio its coordinates back to the base computer at regular intervals, and you could watch it trace out its motion on the screen. At a more practical level, the brain software could correlate the position of the robot with various sensory events and record this information in a detailed history.

One method of recording this history is to grid the universe around the robot and drop some form of marker into that grid for each significant event in the life of the robot. An example of this is shown in *Figure 11-8.* Most significant events in the robot's life consist of physical contact with objects (or, later, sonar information as it scans the environment). Each contact the robot makes indicates the location of some obstacle, temporary (like the family cat) or permanent (like a wall).

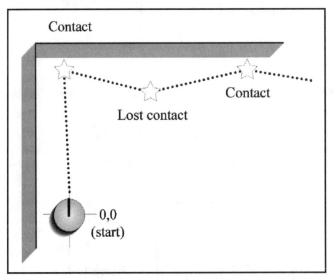

Figure 11-8. Event mapping.

The very act of fitting the world into a grid adds error to the information. Every event or contact is forced into a rigid grid position, changing its location slightly (give or take the grid's dimensions). The more detailed the grid, the better the placement. Except that large grids take huge amounts of memory. Dynamic gridding methods such as quad-tree coding (described in most computer graphics texts) can alleviate the memory problem, at the expense of adding complexity.

As the robot travels further from its origin point, the less precise the position data becomes. Instead of knowing where it is at a precise point in space, the robot's possible position is a fuzzy blob which grows the farther the robot travels *(Figure 11-9)*. This fuzziness could be captured in the robot's internal map by recording events not as a single point, but as a cloud of uncertainty with a size and strength of conviction based on how uncertain the robot's position is. This uncertainty is based on how long it has been since the robot has absolutely located itself.

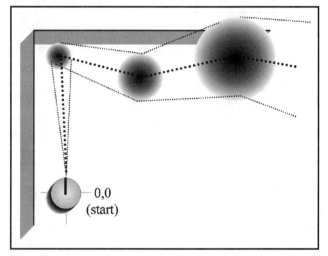

Figure 11-9. Position error over time.

If the robot has a good map of the environment (either predefined, or created over time), it might be able to travel back to some landmark and reorient itself on the map with its sensors. For example, it might travel along a wall until it reaches a known doorway, at which time it knows both its orientation and absolute position in space. This position can be used to determine the relative error of the robot's calculated vs. actual position. If the event and position stream was kept in raw vector format in addition to the gridding, the error term can be propagated back through the sensor history adjusting the grid record as it goes, much like back propagation in neural networks. Once repositioned, the robot can launch out and continue exploring its world, adding to its map and clarifying fuzzy areas.

These tasks of mapping, managing uncertainty, and identifying features on the map given sensor data, are all very difficult, though active, areas of research.

Where am I going? Virtual Senses

Given a sensor created map of the world, the robot can use that map to locate itself in the world. For example, if the robot were picked up and placed in a new room it could (theoretically) determine which room it is in by traversing the walls and matching the landmarks it finds with the landmarks of known rooms from its memory. Possible, though not easy.

With this map the robot can supplement its vague and error-prone physical sensors with virtual sensors inside the virtual world. These sensors can give the robot both fine and coarse awareness of obstacles and information in this virtual map world. This robot ESP can be used for navigation and decision making in the same way as the physical sensors. In fact, the robot's brain may make little or no distinction between real and virtual sensors.

Maps give the robot the ability to plan paths, finding a clear way from point A to point B. Once the path planning part of the robot's brain has calculated an optimal route, it must guide the robot along it. If the lower levels of the robot's intelligence are using a behavioral or subsumption architecture, you can add a map sense. This sense exerts subtle "pressure" on the system, directing it to drive in a particular direction. The other layers of reactive control are still active, avoiding obstacles and servicing higher priority issues.

What's out there? Physical Senses

There are many attributes of the world that a robot may want to sense. Here is a survey of some sensors available to the experimenter, with specific examples. The information returned by most of these senses (or sense modules) is either digital information, a resistance change, or a voltage level. A robot's circuits or MCU can read this information directly, or the signal can be amplified and decoded with an A/D converter.

Optical Feelers

The eyes in previous projects have been entirely passive. They collect and compare the ambient light, and rely on mechanical feelers to sense obstacles. The robot can take an active role in sensing light by sending out its own illumination with an LED or infrared-LED. It can detect the reflection of this light off of obstacles in the environment.

Infrared (IR) is often used for optical feelers since it less likely to become washed-out by ambient light, unless the robot operates outside or in direct sunlight. Infrared LEDs can be easily found, as well as matching infrared photodetectors. The hard part with IR is testing, since the light is invisible to human eyes. Radio Shack carries a little plastic card that makes it visible, and this is handy for testing the output of the IR LEDs, or other IR devices.

Placing an IR LED next to an IR photodetector will saturate the detector whether an object is nearby to reflect the light or not. It is necessary to place the detector into an opaque tube to shield it from the nearby emitter. This tube also acts to limit the field of view of the detector. Longer tubes give a narrower view. A tube can also be placed on the LED to focus its output.

Tubes of the correct size can be made by wrapping opaque tape, sticky side out, around the barrel of a pen. This tape tube can be placed over the LED or detector, and then stuck to the robot's platform.

Feelers may be pointed outward from the robot, or they can be pointed down at the ground. Down-pointing detectors can sense drop-offs (such as stairs), or changes in surface reflectivity so that IR-absorbing black tape or paint can be followed or avoided.

Sharp makes an all-in-one distance sensor (the GP2D05) that signals when an object is within an adjustable distance, though this is an expensive solution to a fairly simple problem.

Modulated IR

Even in the best of environments, there can be stray IR to confuse the robot's optical feelers. One method to avoid false signals is to pulse the light at a specific frequency, and then detect reflected light only at that frequency. Anything else, such as ambient light, is ignored.

Sharp makes a variety of modulated IR detection units designed for remote control systems in home electronics. Radio Shack carries one of these modules, the GP1U58. This unit outputs a digital signal indicating when it is receiving light modulated at the correct frequency, and in this case, it's 40 kHz. LiteON makes similar modules, the LTM-97 series.

An important, and often overlooked, detail when using the Sharp modules is to ground the metal case. If the case is not grounded, electrical noise can cause false signals from the module.

The detector module is designed to detect short bursts of light. 600 uS bursts of light separated by 600 uS of dark is indicated in the specification. This on/off pattern is good for the robot since it can determine if the signal is true or not by ensuring the module turns off when the light is off, and the module turns on when the light goes on. The light itself can be easily modulated with a 555 timer.

TV Remote Control

An enhancement of simple modulated IR is to overlay the modulated light with coded information like that of the home electronics remote control. The IR Module described above decodes the remote control's carrier frequency and passes binary signals to the MCU. The MCU must then decode those signals and make sense of them.

Teaching the robot to understand TV remote signals is perhaps the easiest way to communicate with it. The challenge lies in the coding of the signal since different manufacturers use not only different bit sequences to represent the same meaning (such as Play or Channel-Up), but they can use entirely different bit coding methods *(Figure 11-10)*.

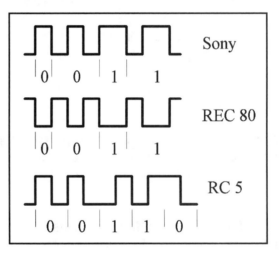

Figure 11-10. Remote control coding.

There are three methods used to code the signal. High-coded (used by Sony), low-coded (REC 80), and phase or transition-coded (RC 5). Different remotes also use different carrier frequencies. For example, Sharp makes GP1U58x series modules to detect carriers at 32.75, 36, 36.7, 38, 40, or 56.8 KHz.

The Sony coding method changes the length of the light pulses, using short pulses for zero and long pulses for one. These are separated by constant duration dark spaces. REC 80 is the opposite, measuring the duration of spaces separated by constant duration light pulses. Both the Sony method and RECS 80 use a different amount of time to transmit a 0 or a 1 value.

RC 5 uses a constant time span for each bit, and it does not measure the duration of either pulses or spaces. Instead, RC 5 coding detects a transition during the bit's time slot. A high-to-low transition indicates a zero, and a low-to-high transition indicates a one.

You can experiment with the remotes around your house, or find the detailed specifications for a remote from the manufacturer or the internet.

Distance

Reflected light can be used to detect not only the presence of an object, but its distance from the detector *(Figure 11-11)*.

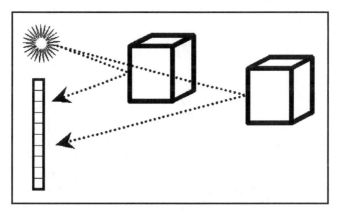

Figure 11-11. Optical distance measurement.

Light is emitted from a distance-measuring sensor (such as the Sharp GP2D02 or GP2D12) and reflected by an object. This reflection is focused onto a linear array of sensors, and the distance of the object is indicated by where the reflection falls on this array. After suitable processing, the distance value is sent to the MCU for use. The GP2D02 communicates with the MCU through a simple serial interface.

The Sharp sensors are supposed to be immune to variations in color or reflectivity in the objects it senses, and can measure distances from 10 to 80 centimeters (roughly 4 to 30 inches).

Sonar

While distance is measured with light by measuring the position of a reflection on a sensor, distance is measured with sound by measuring the time it takes for a pulse of sound to reach an object and return to the sensor. Though it is possible to measure the time-of-flight of light to get extremely accurate distance measurements, it is beyond the capability (and bank balance) of most experimenters.

Sonar can be performed with a single sensor such as the Polaroid 6500 Sonar system, with its electrostatic sensor, or with a matched pair of transducers. Transmission and reception of a sound signal is similar to sending and receiving modulated light, with a few added complications.

As the signal propagates out from the transmitter, it spreads and weakens. Once it strikes an object, only a portion of the signal is reflected back in the direction of the receiver *(Figure 11-12)*. Both the distance and the size of the object affect the strength of the return signal. Immediately after sending the sonar "chirp", the receiving circuit must be disabled so it does not report the zero-distance measurement from the chirp itself. As time goes by, the receiver must increase the sensitivity of its amplifier to compensate for the lack of signal intensity over distance and time.

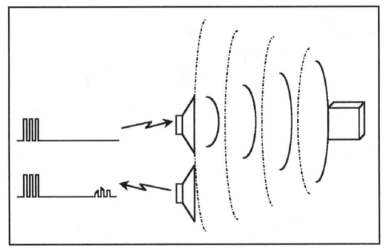

Figure 11-12. Sonar distance measurement.

While mirrors and shiny surfaces may confound light measuring systems, sonar reflects off of any smooth, hard surface. For example, walls are invisible to sonar if they are at a sharp angle to the sensor. Inside corners can also create complex echoes that give confusing readings.

Heat

There are two approaches to sensing heat. The first uses a pyroelectric sensor (like the Eltec 442-3) to detect low-infrared radiation, useful in robots for finding humans and fires. This is like the photodetector eyes, but is used to detect or track the source of heat. The heat signal from humans is very weak, so a special lens is used to collect and focus it on the sensor. The detector and the lens can usually be found together from the same supplier.

The other method of sensing heat is used to detect the ambient temperature, and this requires a different type of sensor such as the National Semiconductor LM335. These devices are like many other sensors and generate a voltage that varies with the attribute being measured. In this case, heat.

An even more flexible device is Dallas Semiconductor's DS1629 digital thermometer and real time clock. This device is designed for use as a home thermostat controller. It communicates with the MCU through a serial interface, and provides accurate time and temperature readings.

Stretch and Bend

Not all sensors are designed to measure such ephemera as light, sound, or heat. It is often necessary to quantify very tangible attributes. The Merlin stretch sensor is an innovative variable resistor in the shape of a round, black rubber band. It is available in a variety of lengths. The resistance through the sensor increases linearly as a function of the band's length. The optimum, repeatable stretch of this sensor is 50% of its base size, though it can operate at up to 100% extension and is rated to break at 300% stretch.

This sensor is an inexpensive replacement for linear potentiometers, and is small enough to be useful in most applications. The sensor is conductive along its entire length, so it is important to keep it electrically isolated from its environment. Also, extreme vibration of the sensor may give some anomalous readings.

Images Company also makes a unique sensor which measures the amount of bend or curvature. This 4-inch sensor can be used to detect joint angles in arms, or as part of clothing on an operator to provide control signals for teleoperation. Like the stretch sensor, it returns its information in the form of increased resistance as the sensor bends.

Acceleration and Orientation

In mammals, the inner ear system provides vital information to their brains about vertical orientation (balance) and acceleration. This information is so central to our physical operation that subtle discrepancies can throw our bodies into a tizzy, i.e. motion sickness.

Analog Devices makes a sensor that can detect either AC accelerations (such as vibration) or DC accelerations (such as inertial force and gravity). The ADXL05 is very sensitive, and the ADXL50 can detect much larger accelerations.

These sensors, and their relatives, are used in inertial guidance systems. By accumulating the intensity and direction of force over time it is possible to keep track of your linear position, though rotation can be a problem. A set of sensors can be used, with appropriate math, to sense the accelerations inherent in rotation but there are better sensors to manage that task.

A sense of gravity is useful for any walking robot because its gives it a sense of balance. Rolling robots can also use a sense of balance to measure the inclination of the surface.

Precision Navigation makes some capable compass sensor modules (such as the TCM2 and Vector2X series), some of which also incorporate a sense of gravity. Though somewhat expensive, they have a simple serial interface and provide an accurate internal compass which is useful during navigation.

Pressure

Though a sense of air pressure may not have as obvious an application as a sense of direction or light, Motorola makes a variety of pressure sensors in their MPX line of products, such as the MPX2700 sensor. Air pressure can be detected against a fixed reference pressure, relative to the ambient air pressure, or as the differential pressure of two ports.

A sense of touch can be made by attaching soft rubber tubing or bladders to a robot's perimeter or fingers and piping the air from this to the sensor. Any contact will change the air pressure in the sensor. The MCU can measure the extent of that pressure for force feedback, or simply switch on a pressure change for bump sensing.

Higher-pressure sensors can be used in pneumatic systems to provide internal force feedback.

A cleaning robot may want to sense the suction in its vacuum system, to determine when the bag is full, or if the inlet is clogged.

Battery Voltage-Level Sensor

An important internal sense for a robot is a sense of power. It needs to know how much time it has left to live on the current batteries. The voltage of a battery drops in a predictable manner over its lifetime, though this decay curve is different for different types of battery. An analog-to-digital converter can be used to measure this voltage.

The challenge is in reducing the battery voltage to the range recognized by the A/D system. The converter needs the measured voltage to lie between the two extremes of zero (ground) and the supply voltage of the converter. Battery voltage begins at, for example, 9-volts and drops until such time as the voltage regulator can no longer sustain the circuits. A simple resistor voltage divider can be used to reduce the battery voltage by some constant factor, such as half. The A/D converter simply measures this reduced voltage. Internally, the MCU can multiply the value back out to get the actual voltage, or it can operate directly with the factored value.

CHAPTER 12
R/C Servos

Standard radio control servomotors that are used in model airplanes, cars, and boats are useful for making arms, legs, and other mechanical appendages that move back and forth rather than rotate in circles. With a little modification, these servos can also serve as rotary motors like the ones described in earlier chapters. In fact, R/C servos can rotate forwards or backwards depending on the width of the control pulse on their one control line.

A generic servo is shown in *Figure 12-1*. A variety of levers and discs can be mounted on the toothed axle protruding up from the servo's box.

Figure 12-1. R/C servo.

After removing the four long screws from the bottom of the servo, the top and bottom covers can be easily removed. This exposes the gears on the top, and the circuit board on the bottom *(Figure 12-2)*.

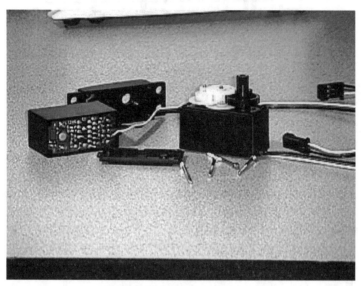

Figure 12-2. Opened servo.

The gears are easily removed, though you need to remember how they were originally placed in the servo. The motor's drive gear is to the left of the servo, where it normally engages with the other gears. To the right is the servo's feedback potentiometer shaft that normally sits inside the output gear *(Figure 12-3)*.

Figure 12-3. Servo gears.

The output gear normally has a small plastic stop to prevent it from rotating too far in either direction *(Figure 12-4)*. The first modification is to carefully cut that stop off so the output gear is smooth and flat along its entire surface. Servo intervals are shown in *Figure 12-5*.

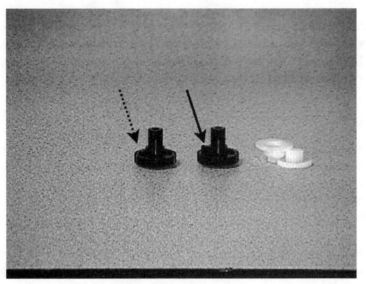

Figure 12-4. Gear stop.

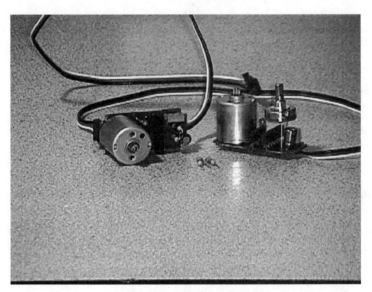

Figure 12-5. Servo intervals.

Press the potentiometer shaft against a table or other hard surface to pop the circuit board out of the case. You may also need to carefully pry the corners with a small screwdriver or other slender object to remove the circuit board holding the motor and control logic.

On this assembly is the feedback potentiometer. The second modification is to re-move this potentiometer and replace it with a fixed resistor network. First, clip the leads of the potentiometer, and then desolder and remove the orphaned leads from the circuit board.

Next, take two matched resistors whose sum is about the same as the value of the potentiometer. In this servo the potentiometer is 5 Kohms, so I used two 2.2K resis-tors. Solder these resistors together so they make a three-lead device. Solder this potentiometer replacement into the servo circuit.

Finally, reassemble all of the pieces.

Driving the servos is fairly simple. On the Futaba-style servo, the red wire is con-nected to power (about 6 volts), the black wire to ground, and the white wire to a digital control signal. The control signal centers the servo when it is around 1.5 mS wide and changes the servo position to the left and right for narrower and wider signals. Each signal should be no more than about 20 mS apart.

The code to drive eight servos is shown in Servo.fuz. The servo position is taken from the Analog to Digital converter, and the signals are generated on the Digital Out port. Each servo is positioned in 128 steps with control values from 0 to 255.

```
define Servo
{
        time as byte[16];
        beat as byte;
        port as byte;
        signal as byte;

        do Init();                               // Initialize Heartbeat timer

        do Pulse() as interrupt      7;          // Interrupt handler on Timer-0
           const ON = 0x10;
}
```

The servo object does very little beyond defining the control data. The time array carries the pulse start and stop delays for each servo channel. The beat variable steps through this array at each Pulse interrupt, and port holds the current servo channel port number. Signal hold the value of the digital out port since the port is write-only and cannot be used as a reference when manipulating its bits.

```
Servo:Pulse()
{
        System:TCNT0 = 255 - time[beat];
        if (beat & 0x01)
        {
                signal = signal & ~port;
                port = port << 1;
        }
```

```
        else
        {
                signal  = signal | port;
        }
        ioPort:DigitalOut = signal;
        beat = beat + 1;

        if (beat >= 16)
        {
                idx as byte;
                val as byte;

                beat = 0;
                System:PORTD = System:PORTD ^ ON;

                port = 0;
                idx = 0;
                while (idx < 16)
                {
                        val = ioPort:Analog[port] >> 1;

                        time[idx]  = 128 + val;
                        time[idx+1] = 32 + (128 - val);

                        port = port + 1;
                        idx = idx + 2;
                }
                port = 1;
        }
}
```

The Pulse interrupt handler is called every 1 to 2 mS or so, depending on values in the control array. Every time Pulse is called, one of the servo control bits is either set or cleared. The first time through, beat is zero, so port bit 1 is set and the timer is set to interrupt the system again when it is time to clear that signal.

The second time Pulse is called, beat is one, so the port is cleared and the port counter is incremented to the next port. The timer is set to delay a small amount until the next port is to be set. These spaces between pulses on the different ports keeps each servo control signal with a consistent repeat frequency.

The third call sees beat with a value of two, which is low-bit cleared so the next port is set. The cycle continues for all eight ports, or sixteen interrupts.

After the sixteenth interrupt is managed, the control array has its delay values reset with the information taken from the analog-to-digital converter. This converter is always running in the background on its own interrupt, so the values in this array are always up to date. Once the control array is filled, beat and port are set back to the beginning and the cycle starts over again.

With eight servo ports available, the robot could have a pair of wheels, an arm, some scanning light or sound detectors, or many other peripherals. It should be possible to make a legged robot, or some other unusual configuration.

However, note that the servos draw a lot of power – and more when they are heavily loaded. The servos should be driven from a separate 5 or 6-volt power supply, the same way the drive motors on the robot platform have their own batteries. When running the eight-servo test circuit, I had to attach a 1-amp power supply before they would work correctly.

With many servos on one power system, any brownouts cause neighboring servos to "twitch", which may cause further brownout. This can cause more servo control failures until the entire array is jiggling uncontrollably. If your servo are responding poorly, check to see if the control signal is bad or the power supply is failing.

CHAPTER 13
Pneumatics

Though there are few small robotic systems using pneumatics, a large number of commercial robots use fluid power in one form or another. Pneumatics is also used in a variety of animatronic systems. Pneumatics holds a personal interest for the author since air power is the basis of Project Boris, a large six-legged mobile robot .

Poetically, using pneumatic power under electrical control brings forth the full power of the storm. Wind and lightening, twisting and surging under the supervision of software written by our own mortal hands.

Practically speaking, air power gives a strength, speed, and grace of motion that is hard to achieve with other types of actuators. The largest drawback to pneumatics is their requirement for compressed air. Compressed air is usually generated by large, stationary, noisy air compressors.

Pneumatics falls under the category of fluid power, and the other branch of fluid power is hydraulics. The "hydro" in hydraulics does not refer to water, but to a special oil used in the system. Hydraulics is the most powerful actuation system available for machines, and it provides precise and stable motion. Hydraulic actuators drive most of the heavy machinery in the world, and is found disguised as the brake system in your car. The problem with hydraulics for the experimenter is in the oil. The oil required to run the system adds considerable weight to the system, and it is extremely messy. Hydraulic oil is not suitable for any dining room laboratory, and may not even be a welcome guest in most garages.

Air power has its own unique challenges once you find a way past the air-compressor problem. The following sections introduce air power, and review some of the ways to tame it economically.

Air Power Principles

While this book can assume the average reader has some experience with electronic circuits, it can't assume a familiarity with pneumatics. Basic terms, values, and formulas are introduced and defined here.

Air Pressure, Force, Volume, and Temperature

Pressure

Air is composed of many tiny molecules zipping around at a high speed, banging against the edges of whatever objects are in their path. Squeeze extra molecules into a closed container and you have more molecules banging into the inside walls than the outside. This banging translates into a force on the inside of the container.

The air around us on Earth, at sea level, is pressing against us (inside and out) with 14.69595 pounds of force for every square inch of exposed surface. As you go above sea level, this pressure reduces until you reach the vacuum of space.

Air pressure is typically measured in pounds per square inch, at least in America. The constant value of 14.7 pounds per square inch is referred to as 1 "atmosphere". Other units of air pressure include the bar, Pascal (or more commonly, kilopascal), and inches of mercury. Actually, inches of mercury is one you don't often see in pneumatics.

    1 kilopascal = 0.1450377 PSI
    1 atmosphere = 14.69595 PSI
    1 bar = 14.50377 PSI
    1 inch of mercury = 0.4911541 PSI

When measuring air pressure, you can compare it to the current atmospheric pressure (gauge pressure), or to a vacuum (absolute pressure). Absolute pressure is the gauge pressure plus 14.7.

Force

To be useful, air pressure must be converted into a mechanical force. Force is a factor of the air pressure (in PSI) and the surface area it is applied to.

$$Force_{pounds} = Pressure_{PSI} \times Area_{inches}$$

As either the pressure or the area increases, so does the applied force. This is why large diameter air cylinders push harder than smaller cylinders. Of course, moving a larger piston a given distance takes more air than moving a small piston that same distance so the larger cylinder moves proportionally slower.

Pressure, Volume, and Temperature

There is a fixed relationship between air's pressure, volume, and temperature.

P = absolute pressure in pounds per square inch
V = volume in cubic inches
T = absolute temperature in degrees Rankine ($T_{Rankine} = T_{Fahrenheit} + 459.67$)

Boyle's law says that the volume and pressure of a fixed amount of air are inversely proportional to each other. This means if you squeeze that air into a smaller space, the pressure increases. If the pressure is increased, the air fits into a smaller space:

$$P_1 \times V_1 = P_2 \times V_2$$

The trouble with this formula is that it is only correct when the air is the same temperature on both sides of the equality. However, as described by Charle's Law, the temperature of a given amount of gas changes as its volume changes. Increase the volume and the temperature falls; squeeze it smaller and the temperature rises:

$$V_1 / T_1 = V_2 / T_2$$

Gay-Lussac relates temperature to pressure, an obvious next step considering Boyle's law:

$$P_1 / T_1 = P_2 / T_2$$

Finally, all of these laws combine into a general Gas Law:

$$P_1 \times V_1 / T_1 = P_2 \times V_2 / T_2$$

With this equation you can get an idea of how a change in one attribute of a gas affects the other factors. For example, raising the pressure of the gas in a fixed container will increase the temperature.

Standard Air

For purposes of comparison, industry has a "gold standard" for air:

Pressure at 14.7 PSI absolute
Temperature at 20 degrees Celsius (20 Celsius = 68 degrees Fahrenheit, or 528 Rankine)
Humidity at 36 percent

Air Flow

Using air in a system implies moving that air from one place to another in the circuit. Air, like electricity, encounters resistance in its travels. Air, like current, flows at a fixed rate through that resistance at a given pressure.

There are two common metrics for air flow, C_V and SCFM.

SCFM stands for Standard Cubic Feet per Minute, which is one cubic foot of standard air transferred in one minute. A cubic foot is 12x12x12 = 1,728 cubic inches.

C_V is a measure of a device's ability to pass air, or its flow coefficient. C_V is useful for comparing valve capacity. The air flow is affected by many things in the circuit, including the pressure at the inlet and outlet orifice of the circuit being tested; the orifice size and shape; the length of any tubes carrying the air; turbulence caused by bends in that pipe; and any restrictions in the flow path.

C_V is often directly measured, though it helps to understand what it represents. It incorporates many attributes of the system into one handy value: the flow metric of SCFM, the pressure at the inlet relative to the pressure at the outlet, and the temperature and specific gravity of the fluid (in our case, air). Another important aspect of any valve (or other pneumatic circuit) is the pressure drop across the circuit, which is determined by some of the same factors that define C_V.

> Q= SCFM = Air flow of standard air, in cubic feet per minute
> P_1 = Absolute pressure at the inlet
> P_2 = Absolute pressure at the outlet
> G = 1 (specific gravity of air)
> T = Absolute temperature ($T = T_{Fahrenheit} + 460$)
> $C_V = (Q / 22.48) \times \mathrm{sqrt}((G \times T) / ((P_1 - P_2) \times P_2))$

The formula to determine the air flow required to move a pneumatic cylinder a specific amount is:

> P = Pressure in the cylinder in PSI
> A = Effective surface area of the cylinder's piston in square inches
> D = Cylinder stroke distance in inches
> N = Number of cylinder strokes per minute
> Q = SCFM = $(A \times D \times (1 + (P / 14.7)) \times N) / 1728$

The C_V required for this cylinder can be calculated using the C_V formula above, using a standard 5 of 10 PSI pressure drop for reference.

For example, calculate the air needs of a cylinder with a 2-inch diameter cylinder with a 6-inch stroke, operating at 100 PSI with a circuit pressure drop of 10 PSI. Say this cylinder must cycle 10 times per minute (5 in, 5 out), or 6 seconds per stroke.

$$P = 100$$
$$A = 2 \times \pi = 6.28$$
$$D = 6$$
$$N = 10$$
$$Q = (6.28 \times 6 \times (1 + (100 / 14.7)) \times 10) / 1728 = 1.70 \text{ SCFM}$$

The C_v required to support this airflow is then:

$$P_1 = P = 100$$
$$P_2 = P - 10 = 90$$
$$G = 1$$
$$T = 520$$
$$C_v = (Q / 22.48) \times \text{sqrt}((G \times T) / ((P_1 - P_2) \times P_2))$$
$$Cv = (1.70 / 22.48) \times \text{sqrt}(520 / (10 \times 90)) = .057$$

To drive this circuit, find a valve with a C_v of .057 or greater.

Actuator Force

The linear force applied by a cylinder is much easier to determine than its airflow requirements.

The geometry of a standard air cylinder is such that the extension force is greater than the retraction force. Refer to *Figure 13-1* for the analysis that follows.

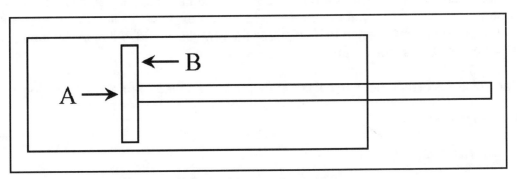

Figure 13-1. Pneumatic cylinder.

The effective surface area of the cylinder at A is a simple function of the cylinder's diameter D_C:

$$A = D_C \times \pi$$

The effective surface area of the cylinder at B, the side used during retraction, must account for the diameter of the piston's rod:

$$B = (D_C - D_R) \times \pi$$

This, of course, assumes the cylinder is using air pressure for both extension and retraction. Many cylinders only use air in one direction and use springs to return to their starting position.

The force exerted is a simple function of the air pressure in PSI times the effective surface area of the cylinder:

$$F = P \times A$$

A subtle and extremely important modifier on the cylinders thrust is whether the force is measured in motion or at rest and fully pressurized. A pneumatic cylinder generates its maximum force when it is at rest and at full pressure. When the piston in the cylinder is in motion, the cylinder generates roughly half of its full force. If the cylinder encounters resistance while it is in motion, it will slow significantly to allow the pressure to build behind the piston until that resistance is overcome.

Components

Electrical circuits have a power supply, components to manipulate the electricity in the form of transistors, resistors, capacitors, and integrated circuits, and various input controls. Robots use inputs such as the sensors explored in this book, and add motors or some other device to manipulate the physical world. All of this is then tied together with wires, solder, and PC boards.

Each of these electromechanical components has an electro-pneumatic analog.

Air Supply

Pressurized Air

The air supply provides the force to make the pneumatic system work. The end product of any air supply is a flow of pressurized air. Two principle methods of supplying this pressurized air flow are with an air compressor and with an air storage tank. Air supply symbols are shown in *Figure 13-2*.

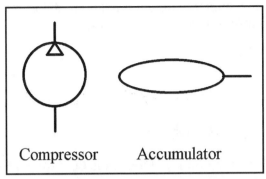

Figure 13-2. Air supply symbols.

Air compressors take ambient air from the environment and stuff it into a container at the time the air is needed. Air compressors have two rating values, air flow in CFM and their working pressure. Often a compressor will specify SCFM at two or more representative pressures. For example, the Coleman Contractor air compressor with the gas-powered 5HP Briggs & Stratton engine displaces 12.0 CFM at 40 PSI and 10.0 CFM at 90 PSI, with a maximum pressure of 125 PSI.

Compressors can be portable or stationary, electric or gas powered. The correct compressor for an application depends on many factors, such as the pressure and CFM required, and any weight, mobility, or noise restrictions.

Many applications can use of a bottle of compressed air or other safe gas, such as carbon dioxide or nitrogen. Compressed gas can be found at welding supply houses and scuba supply stores. Though these tanks are normally made of (heavy) metal, it is also possible to find high-pressure containers using wound composites that are lighter and actually safer.

Whenever you use high pressure gas containers, it is vital to be aware of the dangers involved. A large tank of gas compressed at, say, 6,000 PSI turns into a deadly missile the moment the valve stem breaks off. This should never happen under ordinary use, but placing such a bottle into a robot is far from ordinary.

When using compressed air from a container, there are two issues to resolve. The first is to determine the size and pressure necessary to run the application over the necessary time frame. The second issue is more subtle, and involves the relationship of pressure and volume to temperature. In short, depressurizing air makes it cold, and in some cases cold enough to freeze the moisture in a pneumatic circuit solid.

If your pneumatic circuit uses an average of 2 cubic-feet of 95 PSI air per minute, and it must operate for, say, 10 minutes, it uses a total of 20 cubic feet of 95 PSI air during that operation. If the air containers hold a maximum of, for example, 5,000 PSI then the container must be able to hold V_2 volume of air:

$$P_1 = 95$$
$$V_1 = 20$$
$$P_2 = 5,000$$
$$V_2 \times P_2 = V_1 \times P_1$$
$$V_2 = (V_1 \times P_1) / P_2 = .38 \text{ cubic feet}$$

Of course, if the starting temperature of this air is 60 degrees Fahrenheit (520 absolute), the ending temperature is:

$$T1 = 520$$
$$P_2 \times V_2 / T_2 = P_1 \times V_1 / T_1$$
$$T2 = (P2 \times V2) \times (T1 / (P1 \times V1)) = 494$$

494 degrees absolute is 34 degrees Fahrenheit – getting cold. Of course, in a real system, this drop in temperature causes an additional drop in pressure, which affects the volume, air flow, and each one again in a circle. These simple formulas only give a general estimate of the actual performance of the system.

Regulation

Air from a compressor will usually be 90 to 100 PSI, which is the basic operating pressure of many pneumatic components. Even so, it is useful to place a pressure regulator (with a gauge) between the compressor's air accumulator and the pneumatic circuit. Using the regulator you can test the system at low-pressures, where it is slow and weak, before moving up to full power. Air regulator symbols are shown in *Figure 13-3.*

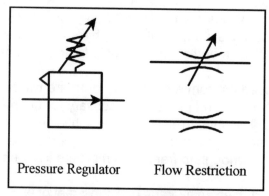

Pressure Regulator   Flow Restriction

Figure 13-3. Air regulator symbols.

The pressure regulator should have a gauge on the low-pressure side to show the working pressure. Regulators for high-pressure tanks have a gauge on both sides, one to show the pressure of the tank and another to give the working pressure.

When using a high-pressure tank, it is good to add a storage container on the low-pressure side as an expansion chamber. This buffers the recently decompressed high-pressure air, and provides a reserve of medium-pressure air for the system to draw on. For some systems, you may even need to place small reservoirs near the actuators to limit pressure drop across the system.

Flow regulators, either fixed or variable, can be added to the circuit at either the air source or at each actuator. Restricting the air flow slows the actuator's motion without reducing the air pressure.

Clean and Dry

Air from a tank is probably already clean, dry, and ready to use. Air from a compressor, however, is going to be damp and dirty. Pneumatic valves and cylinders are precise mechanical devices that react poorly to dirt, and can be damaged by excessive moisture.

The minimum requirement for any pneumatic circuit is an air filter between the compressor and the components. The filter extracts dust and particles from the air and will also trap some of the moisture. Sensitive pneumatic systems may have several stages of cleaning and drying for their air to remove particles, moisture, and oil. Air treatment symbols are shown in *Figure 13-4.*

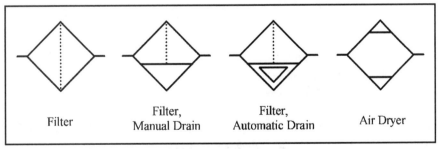

| Filter | Filter, Manual Drain | Filter, Automatic Drain | Air Dryer |

Figure 13-4. Air treatment symbols.

Lubrication

Like any mechanical actuator, pneumatic cylinders and motors need lubrication for optimum efficiency and to reduce wear. Some pneumatic actuators require constant lubrication and this can come from an attachment on the air supply. Other actuators are pre-lubricated "for life" at the factory, though they will benefit from additional lubrication. Too much oil in the air supply could get messy. Each application is unique so discuss your needs with your component supplier. They will know the specific requirements of the components you are using. An air lubricator symbol is shown in *Figure 13-5.*

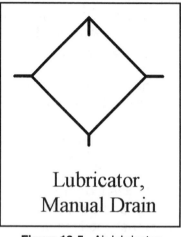

Figure 13-5. Air lubricator symbol.

Air Control

A pressurized air supply is not useful until there is a mechanism to control it—turn it on, turn it off, or redirect it. This air management is provided by pneumatic valves, which are in turn controlled by electricity, manual interaction, or even air.

Valve Circuits

There are two important circuit attributes of a valve: the number of positions the valve can take (typically 2 or 3 positions), and the number of ports on the valve. Valve circuits symbols are shown in *Figure 13-6.*

The number of ports determine the type of circuits that can be formed with a given valve position. Depending on the valve, air may be restricted to flowing a particular direction through any given path, or it may be free to flow either direction through that path. You need to check the specifications of the valve to be sure. With two ports there are only two possible states: both ports connected and both ports isolated.

Three ports add more options. Of these three ports, one may be attached to the actuator, a second to the air supply, and a third can be an exhaust. Other configurations are possible, but this one is common. With this configuration of the ports, there are three useful valve circuits: power to the actuator, exhaust to the actuator (which may indicate a spring-return cycle on a single port cylinder), and all ports blocked.

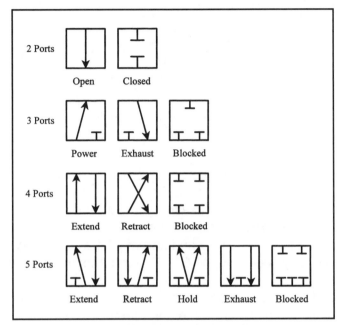

Figure 13-6. Valve circuit symbols.

Four ports can provide pressurized air to two actuator lines for bidirectional control. One circuit in the valve can extend the actuator (or spin it forwards) and another circuit can retract (or reverse) the actuator. The actuator can also be held in position (blocked).

A fifth port can be used to add a second exhaust. A five-port circuit has all of the same behaviors of the four-port circuit, plus the option to open both actuator ports up to an exhaust, releasing all control over the actuator. Some common valve configurations are shown in *Figure 13-7.*

Valves are inherently binary devices. They are either in one position or another, with little gray area in between. Though there are a number of valves (or valve systems) that provide variable flow control instead of simple on/off switching, they are considerably more expensive.

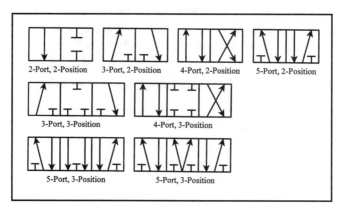

Figure 13-7. Common valves.

Valve Actuators

A valve, having two or more positions, needs a way to switch between those positions. Though the focus of electro-pneumatics is in electrical activation of the valve, there are also mechanically and pneumatically switched valves. In low-voltage electric (solenoid) valves, it is probable that they have an internal pilot line to provide an air boost to the solenoid. Valves with an internal pilot will also have a minimum air pressure requirement, typically around 30 PSI, for that pilot to operate.

Valves may have different actuation mechanisms for each direction. For example, solenoid activation in one direction with a spring return. There may also be multiple actuators for a single valve action, such as solenoid with pilot support, plus a manual override. Valve actuation symbols are shown in *Figure 13-8*.

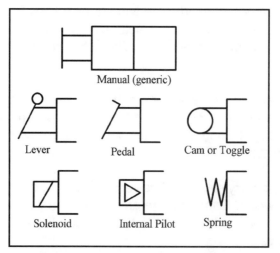

Figure 13-8. Valve actuation symbols.

Solenoid valves provide a way to get around the binary on/off limitation for some valves. Each valve type has a minimum time it takes to switch from on to off, and from off to on. Some valves, such as the Parker B5 series valve, are moderately fast, taking only 10 to 15 mS to perform this switch. Other valves, like the Clippard EC series, are even faster. With this knowledge it is possible to drive the valves with pulse-width or pulse-count modulated signals. The airflow is controlled by the ratio of on to off time, in the same manner as PWM control of electric motors. The pulse frequencies are much lower for valves, and the control isn't always smooth and precise, but it is better than no control at all.

Actuators

There are two basic classes of pneumatic actuator, the cylinder and the rotary vane-style actuator (e.g. air motor). Within these categories there are endless variations, including hybrid mechanisms. There are cylinders that provide rotary motion and vane actuators that create linear motion through gearing systems. Pneumatic actuator symbols are shown in *Figure 13-9*.

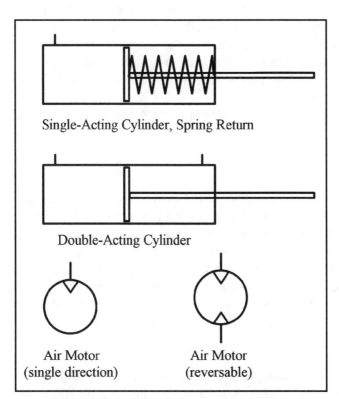

Figure 13-9. Pneumatic actuator symbols.

There is a bewildering array of actuators in the pneumatic industry. The Parker-Hannifin Actuator Products catalog has almost a thousand pages, containing a half-dozen styles of cylinder with dozens of variations each. At the heart of most of these options lie a few simple components. The rest of this chapter discusses the linear cylinder, and ignores the rotary cylinder and rotary air motor actuators.

The body of a cylinder is a simple tube which contains the piston and rod. This tube holds the head and cap into place. The head and cap may be crimped onto the body, or there may be tie-rods bolting the head and cap to either end of the body. The assembly style determines whether the cylinder is disposable or repairable and affects the overall strength and durability of the cylinder.

The head of the cylinder is a machined block that seals one end of the cylinder body. The rod passes through a hole in the head's center. The head contains various bearing and seal assemblies to preserve the airtight integrity of the cylinder. The cap is the block at the other end of the cylinder where the rod does not pass through. Ports are machined into the head and/or cap to supply air to move the piston.

The piston lives inside the cylinder body and has an airtight seal against the interior surface of the body. When air is forced in through a port at one end or another, the increased air pressure on that side of the piston forces it to the other end of the cylinder. A feature in some pistons is a pneumatic "cushion" to slow the piston as it reaches its extreme, saving wear and tear on the entire system.

Attached to the piston is the rod, which in most cases protrudes through the head at one end of the cylinder. The rod carries the motion of the piston out of the cylinder.

Connectors

Tying the entire pneumatic system together is a system of tubes and connectors. The size and nature of the tubing depends on the system. Connectors come in many flavors. Quick connectors as shown in the diagram allow pneumatic subsystems to be easily detached from the whole. Everywhere tubing attaches to another component, there will be some form of connector: straight, elbow, 'T'. These can be quick-release or clamped-in. A well-stocked pneumatic supply shop is like a candy store, full of colorful catalogs and exciting bits of metal and plastic. I've found that the people who work there are sympathetic to robotics projects and are more than willing to help.

Most actuators have a simple threaded port for the fitting where the supply tube attaches. Valves may attach directly to the air lines (in-line). Valves may also be bolted together on a common base (manifold) with one air supply, an exhaust or two, and then control lines to each device. Threaded fittings and in-line connection systems are fairly generic across all devices. Valve manifolds are created by the manufacturer for a specific family of valves. Line and connector symbols are shown in *Figure 13-10*.

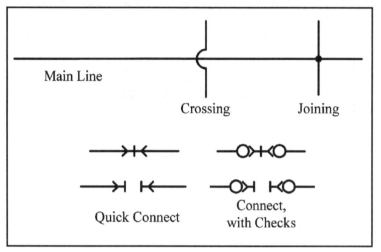

Figure 13-10. Line and connector symbols.

Feedback

Typical pneumatic cylinders do not have any sense of their position, and they are typically held in only two positions, fully retracted and fully extended. A surprisingly expensive add-on places a magnet on the piston and a reed-switch (or hall-effect sensor) at one or both ends of the cylinder to provide feedback at the piston's extreme positions. Extremely expensive cylinders can also provide detailed position information through linear potentiometers within the cylinder, or even sonar-based position sensing.

There are two systems that can provide feedback to a computer controller: the air system and the mechanical devices attached to it.

Assuming a strictly external set of sensors (nothing built-in to the valves or actuators), the pneumatic circuit has only one attribute to sense, and that is its pressure. The pressure can be detected at the air supply, at the valve, and at the cylinder. How you measure the mechanical system's position depends on what type of mechanism it is, and can take a lot of ingenuity.

Mechanical Issues

The cylinder allows for several types of attachment. The rod is usually threaded at one end, so it can be bolted directly to a mechanism. Also, a clevis can be attached, providing a pivot joint. The cylinder's head or body may have one of several types of mechanical joint. The head may be threaded for a rigid nose mount, have threaded holes for a front block mount, or have projections for trunnion mounting. The end cap may have any of these mounting forms or a rear pivot mount *(See Figure 13-11).*

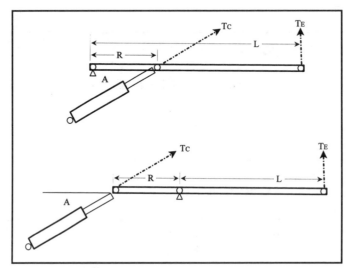

Figure 13-11. Mechanical system.

The force exerted to a given task by a cylinder depends on the thrust the cylinder develops in combination with the geometry of the mechanical system.

The cylinder exerts a linear thrust T_C which is applied to a rigid lever. The lever and the cylinder both have pivot mountings. Whether the lever has the fulcrum at its end and the cylinder is attached in the center, or the fulcrum is at the center with the cylinder at the end, the math is essentially the same.

The angle of the cylinder A with respect to the lever determines how T_C is applied to the load. The force directly perpendicular to the lever is sine(T_C), and this force component does all of the work. The force along the lever is cosine(T_C), and this force component only serves to increase the friction and stress in the system.

The leverage of the system can increase or, more likely, decrease the effective thrust at the end of the lever. The final effective thrust, is:

$$T_E = (T_C \times sine(A) \times R) / L$$

As the cylinder is extended or retracted, the angle changes and so does the effective force. As A approaches zero, not only is the efficiency of the mechanism lost but there is a real possibility that it can jam or reverse. It is important that A is never very close to either 0 or 180 degrees.

Appendix A: Fuzbol Language Reference

Fuzbol is a "small" language dedicated to hardware control. It features specialized fuzzy-logic data types and operators. It currently exists only for the Atmel AT90S8515 processor, though it would be simple to port to other microcontrollers. Though significantly slower in execution than machine language, this loss of speed is often more than compensated by the ease of development a high-level language gives.

Each Fuzbol statement is transformed into one or more byte-encoded instructions by the compiler. These codes are executed by the MCU-based Fuzbol interpreter. Each instruction uses a minimum of 16 bits of RAM, and some are significantly larger. Each execution step has an overhead of six or eight machine cycles (which is good for an interpreter), plus the duration of the working code (which is fairly constant for either an interpreter, compiler, or hand-built code). If the instruction codes are stored in on-chip RAM (for small programs), the overhead is six cycles. The extra two cycles for off-chip RAM are imposed by the Atmel architecture.

Interrupts are disabled during the execution of the Fuzbol start() program, so all hardware interfaces used during the execution of start() must use flag polling techniques. Once start() has ended, interrupts are enabled and the MCU is put to sleep so that any interrupts activated during start() can trigger Fuzbol handlers.

Fuzbol programs may have a polling architecture, or they may be interrupt driven, but they may not use a mixed model. This limitation was imposed to reduce the overhead needed for each instruction step, and to make the execution of interrupts more predictable.

An interrupt may have up to a four cycle delay on its startup imposed by the Atmel architecture. Fuzbol adds 16 to 18 cycles vectoring to the interrupt handler. The time taken by the handler function depends on its contents, though all handlers should be short since interrupts are disabled during their execution.

Further details and program updates can be found at:

http://www.simreal.com/Fuzbol

This reference is for Fuzbol version 1.1.

Hardware Assumptions

Fuzbol makes several assumptions about the hardware it is running on. These assumptions are detailed here.

1. Fuzbol currently only runs on the Atmel AT90S8515 MCU, and it assumes an 8 MHz clock.

2. Port-B, Pin 2 is configured as an input (with pull-up active) on reset. If this pin is allowed to remain high on reset, the Fuzbol program loader is entered. The loader then waits for a program to be sent to it. If Port B Pin 2 is pulled low during reset, any Fuzbol program in memory is executed.

3. Port B, Pins 4 through 7 are used by the SPI port. This port is dual purpose. Fuzbol itself is loaded into the chip's flash memory through the port. Fuzbol code is also loaded into static RAM by the Fuzbol loader through the SPI.

4. Port D, Pin 4 is configured as an output and then set to on during power-up reset. Any Fuzbol program can override this behavior as needed.

5. Port D, Pins 0 and 1 are used by the UART. On reset, Fuzbol configures the UART to 9600 baud, eight bits, 1 stop, no parity. Fuzbol code can override these settings for its own use.

6. The RESET pin is also wired into the SPI connection. The in-circuit programmer must have control over reset to program flash memory. Reset should be wired to Vcc by way of a 1Kohm resistor, to avoid conflict with the programmer.

7. Fuzbol supports non-volatile external SRAM. All startup and interrupt vectors are preserved in EEPROM.

Fuzbol Programs

```
<program> := <program statement>*

<program statement> := <use>
                     | <object>
                     | <constant>
                     | <range>
                     | <rule>
                     | <method>
```

A Fuzbol program is defined as one or more program statements. A program statement may be a "use" file-include statement, an object definition, a constant value declaration, a range or rule definition, or a program method (which is a function, subroutine, or interrupt handler). These statements (or in the case of <method>, blocks of statements) in turn include other Fuzbol commands. Each of these terms is defined later. First, a detour into some important Fuzbol concepts.

Tokens

Fuzbol reads program source as text from a file. This file consists of tokens, or groups of characters, which are analogous to the words and numbers in human language. The tokens may be numbers, identifiers (names of things), or Fuzbol instructions (reserved words). Tokens are case sensitive, and Fuzbol instructions are defined with lower-case letters.

Token types include:

Integer

```
<integer> := (0..9)*

0
1
1024
```

An integer is composed of one or more decimal digits. Leading zeros are allowed.

Hex

```
<hexint> := '0x' (0..9a..fA..F)*

0x01
0x1f
0x3c
```

A hexadecimal value consists of 2, 3, or 4 base-16 digits (the decimal digits 0 through 9, plus the first 5 letters a through f), preceded by the marker "0x". Hexadecimal numbers are treated as integers.

Real

```
<real> := [0..9]* '.' (0..9)*

.0
1.0
01.34
2.340
```

A real number has zero or more leading digits, a decimal point '.', plus one or more trailing digits. Leading and trailing zeros are allowed. The minimal real number is ".0".

Fuzzy

```
<fuzzy> := <number> '@' <number>

0@0
1@255
2.75@255
```

A fuzzy value is composed of an integer, real, or hex value separated by the ampersand '@' from another integer, real, or hex weight modifier. The meaning and use of fuzzy values is described elsewhere.

String

```
<string> := '"' .* '"'

"I am a string"
""
```

A string is any arbitrary text (zero or more characters) surrounded by double quotes. Strings must be defined on a single line. At this time, there are no escape codes defined for strings, and they may not contain line breaks or quote marks.

Character

```
<char> := '' . ''

'a'
'X'
' '
```

Any single character surrounded by single-quotes. There are no escape codes defined for characters to allow the specification of non-printing characters.

Identifier

```
<identifier> := (a..zA..Z_)[a..zA..Z0..9_]*

Fred
fred
_count1
some_value
```

Identifiers (names) must begin with an alphabetic character or the underbar, and may contain zero or more additional alpha characters, digits, or underbars. Fuzbol is case sensitive.

Statement

In general, a Fuzbol statement is a "line" of code. Statements are typically terminated with a semi-colon ';'. Statements are composed of tokens, and when parsed they define symbols or executable code.

Data Types

The different data types define the units of information storage in the language. Where the tokens define character sequences recognized by the parser, data types attach further meaning to these raw values.

byte

8-bit unsigned integer with a value of 0 to 255.

integer

16-bit signed integer with a value of -32,768 to 32,767

real

24-bit floating-point value. Real numbers are coded with an 8-bit binary exponent from e-126 to e+128, a 1-bit sign flag, and a 15-bit mantissa.

fuzzy

40-bits of information go into fuzzy data. Fuzzy values are stored as a 24-bit fixed point value (16-bit integer plus 8-bit fractional part) plus a 16-bit fixed point weight modifier (8-bit integer plus 8 fraction). Fuzzy numbers and their operations are explained in more detail later.

array

```
buffer as byte[10];
position as real[4];
```

Byte, integer, real, and fuzzy values may be contained in arrays. Arrays are indicated by the square brackets '[' and ']' after the data type. The array size is a constant expression within these brackets that is evaluated at compile time.

memory

```
UDR as memory 0x0a;
```

A memory field is defined by an integer value that names a hardware memory location. The memory address is kept as a 16-bit integer that is used to reference an 8-bit byte. Reads and writes to a memory variable are done as bytes.

Fuzzy Numbers

Fuzzy systems are typically composed of three processes: fuzzification of input data, rules that operate on the fuzzified data, and defuzzification of those results into unique (crisp) values. Most fuzzy systems are dedicated to that task and have no need of "fuzzy numbers". Fuzbol is a language that supports fuzzy operations, among others, and it needs a way to represent fuzzy information between calculations.

In Fuzbol, a fuzzy number has two parts: the value and the weight, separated by the "@" symbol (e.g. 12@255).

When a fuzzy constant is parsed by Fuzbol, it is stored internally as the ratio (value x weight) / weight. The numerator (value x weight) is stored as a 24-bit fixed-point number (16.8), and the denominator (weight) as a 16-bit fixed-point number (8.8).

When crisp numbers are converted to fuzzy values they are assigned a unit weight of 255. When fuzzy values are converted to crisp numbers, the fuzzy ratio is evaluated to get the value represented.

For example:

Fuzzy Constant	Value Conversion	Internal Representation	Defuzzify
2 @ 255	2 * 1	value 2, weight 1	2 / 1 = 2
2 @ 127	2 * 0.5	value 1, weight 0.5	1 / 0.5 = 2

Weights are entered as bytes valued from 0 to 255. Zero indicates no weight, 127 is half (0.5) and 255 is unity (1.0).

Addition is used to accumulate fuzzy values. Multiplication and division do not have any real meaning for fuzzy numbers and are not implemented.

Fuzbol Instructions

'=' – see "Assignment" or "Expression"

As

```
<variable> := <identifier> 'as' <type> ';'


counter as integer;
index as byte;
distance as real;
poke_me as memory 0x100;
code_array as byte[16];
my_function( in_a as integer, in_b as integer) as integer;
```

The as command is used when assigning data types to variables or method return values.

Assignment

```
<assign> := <lhs> '=' <expression> ';'

<lhs> := <field> | (<field> '[' <expression> ']' )

uart:tx = 'A';
sample:buffer[idx] = idx*2;
```

An assignment is defined by two parts. The left-hand side (lhs) and the right-hand side (rhs), associated with the equals sign '='. The lhs indicates a variable or array entry to receive the value. The rhs is an expression whose result is stored in the lhs.

For byte, integer, and real values the lhs takes the result of the expression.

For fuzzy variables the result is combined with the system weight, either qualifying the crisp result of the expression, or factoring the weight of the fuzzy result. The weight-adjusted value is then added to the fuzzy variable; assignment acts as a fuzzy accumulator. The only way to reset a fuzzy variable is to allow it to go out of scope (and become deleted), or to use the clear() instruction.

Assignment to a memory variable places the byte result into the address referenced by the variable. There is no way to redirect a memory variable to reference a different location after it has been declared.

Block

```
<block> := <block statement>*
<block statement> := <constant>
        | <variable>
        | <assign>
        | <out>
        | <call>
        | <loop>
        | <decide>
        | <weight>
        | <clear>
        | '{' <block> '}'
```

A block is a set of zero or more block statements contained within the curly brackets '{' and '}'. Block statements do the work in Fuzbol, and include most of the code-generating commands.

Blocks may be nested. Each nested block has a unique variable scope, allowing it to create local variables, which may overwrite variables of the same name in the larger containing scope. Once the sub-block is exited, any local variables within it are de-allocated and lost.

Call

```
<call> := <identifier> ':' <identifier>

        '(' [<expression> [',' <expression>]* ] ')' ';'

uart:Send();
heading = odom:GetHeading( heading, left_dist, right_dist );
```

A call invokes a method by name. The method must be fully qualified with its object's name, and the call must specify a type-compatible expression for each parameter in the method.

Case – see "If"

Clear

```
<clear> := 'clear' '(' <identifier> ')'

clear(my_fuzzy);
```

All assignments to a fuzzy variable accumulate the assigned value, whose weight is scaled by the system weight. The only way to erase the contents of a fuzzy variable is to use clear() on it, or let it go out of scope and become deleted.

Comment

```
// Ignore all of this text…
```

Comments are proceeded by two forward slashes '//' and continue to the end of the line. Comments are allowed anywhere. There is no form of comment which can span lines, or which can be embedded within a statement.

Constant

```
<constant> := ('const' | 'constant') <identifier> '=' <expression> ';'

const PI = 3.1415926;
constant TWOPI = (2 * PI);
```

Constants create a named symbol to represent a numeric value or expression. Constants do not affect the executable Fuzbol code, but simply define aliases within the compiler.

Constants can be created for integer, real, and fuzzy values. Constants may be defined anywhere in the program, and follow the same scoping rules as other symbols.

Decide – see "If"

Define

```
<object> := ('def' | 'define') <identifier> '{' <object statement>* '}'

<object statement> := <constant>
                    | <variable>
                    | <method prototype>

define my_object
{
        some_value as integer;

        do some_method( parm1 as integer, parm2 as byte);
        do some_interrupt() as interrupt 0x01;
}
```

An object definition requires a unique object identifier and one or more object statements to define the object's content.

Other than the start() procedure, all other variables and methods are grouped within objects. Objects may not be created or destroyed at run time, but are defined by the compiler as the program is parsed.

Fuzbol is designed to mirror the needs of microcontroller hardware. Each device or peripheral should have an object to represent it, defining the access addresses for that device plus any special methods useful in controlling it. Objects can also be used to cluster code and data for other, non-hardware, systems.

Fuzbol is not a "true" object-oriented system. Fuzbol has no inheritance and objects are fully instantiated at compile time. Multiple methods within an object may have the same name so long as they have unique parameter lists.

Each object defines a unique name space. The same variable or method name may be defined within more than one object without any conflict. Objects for many different hardware devices may be used in the same program with little need to worry about overlapping method and data names.

The object definition contains variable, method, and constant declarations. All access to these internal symbols must be qualified by the object name, e.g. "object:symbol".

Do

```
<method prototype> := 'do' <identifier> '(' [<prototype> [',' <prototype>]* ] ')'
        [ 'as' ( <type>) | 'interrupt' <integer> ] ';'

<prototype> := <identifier> 'as' <type>

do my_proc( in1 as integer );
do some_action();
do my_func( in1 as real, in2 as real) as real;
do my_handler() as interrupt 0x03;
```

Method prototypes are declared in an object definition, where their name, prototypes, and return value are specified.

The method's name plus parameter types must be unique within its object, but may overlap methods in other objects. To access any method, it must be fully qualified by the object name. If two methods within an object have the same (case-sensitive) name, they must have disparate parameter lists. A call to a function or procedure will select the best match, performing the minimum parameter type conversions as necessary.

If there is no 'as' component to the method prototype, the method is a procedure and has no return value. Procedures can be called from code blocks but are not valid inside expressions.

If a method has a return type, it is considered a function and can be used in expressions to perform calculations. Functions have a special temporary variable whose data type matches the function's return type, and with the same name as the function. The last value assigned to this variable becomes the function's return value. For example:

```
define math
{
        // …
        do square( value as real ) as real;
}
// …
math:square( value as real )
{
        square = value * value;
}
```

A method with the return type "interrupt" becomes an interrupt handler. The handler is called whenever the interrupt at the specified vector address is triggered. The meaning of these interrupt vectors depend on the processor Fuzbol is targeted for.

The set of valid interrupt vectors for the AT90s8515 are:

Vector Address	Interrupt Description
0x00	Power-on Reset, *(reserved)*
0x01	IRQ-0
0x02	IRQ-1
0x03	Timer1 Capture
0x04	Timer1 Compare (A)
0x05	Timer1 Compare (B)
0x06	Timer1 Overflow
0x07	Timer0 Overflow
0x08	SPI Transfer complete *(reserved)*
0x09	UART Rx Complete
0x0a	UDR Empty
0x0b	UART Tx Complete
0x0c	Analog Comparator

See also "Define".

Else – see "If"

Elseif – see "If"

Expression

```
<expression> := (<expression> <binop> <expression>)
        | (<unop> <expression>)
        | ( '(' <expression> ')' )
        | <call>
        | <number>
        | <fuzzy>
        | <field>
        | (<field> '[' <expression> ']' )

<binop> := … set of binary operators

<unop> := … set of unary operators

<field> := <identifier> [ ':' <identifier> ]

A + B
not A
(A + B) * C
B
123
3@255
object:some_var
my_buffer[ index ]
```

An expression may be a single value, or a complex nest of sub expressions combined by binary and unary operators. Expressions are defined with infix notation and may reference variables, functions, arrays, or constant values.

Most operators work the same across most data types, with some exceptions that are noted as necessary.

When two operands for a binary operator have different data types, the operand with the lesser bit-count is promoted to the type of the greater operand. Operand promotion goes from Byte to Integer to Real to Fuzzy. Since Fuzzy data is not capable of representing all Real values and may not be used with all operators, care must be taken when using Fuzzy types in mixed expressions.

Variables and constants may be used as operands in any expression. Note that expression operators on constant values are evaluated at compile-time.

The various operators are described below. Not all operators apply to all data types.

```
B : byte
I : integer
R : real
F : fuzzy
```

Operator	Symbol	Types	Description
Add	+	B, I, R, F	5 + 3 = 8
Subtract	-	B, I, R, F	5 - 3 = 2
Multiply	*	B, I, R	5 * 3 = 15
Divide	/	B, I, R	5 / 3 = 1 or 1.6667
Modulo	%	B, I	5 % 3 = 2
Unary Minus	-	B, I, R, F	5 + -2 = 3
Bit-AND	&	B, I	5 & 3 = 1
Bit-OR	|	B, I	5 | 3 = 7
Bit-XOR	^	B, I	5 ^ 3 = 4
Bit-NOT	!	B, I	!0x05 = 0xfa
Bool-AND	and	B, I	The result of all Boolean operators
Bool-OR	or	B, I	is a Byte; 0x00 is FALSE, and 0xff
Bool-XOR	xor	B, I	is TRUE. All operands which have a
Bool-NOT	not	B, I	value other than 0 are considered TRUE.
Minimum	min	B, I, R	5 min 3 = 3

		F	5@2 min 4@3 = 5@2 (returns the operand with the minimum weight). A form of fuzzy AND.
Maximum	max	B, I, R	5 max 3 = 5
		F	5@2 max 4@3 = 4@3 A form of fuzzy OR.
Equal	==	B, I, R, F	All equality tests return Boolean
Not-Equal	<>		values of 0x00 or 0xff. With fuzzy
Greater than	>		values, the weights, and not the
Greater or Equal	>=		actual de-fuzzified value determine
Less than	<		all equality results (as
Lesser or Equal	<=		demonstrated in min and max).
Left-shift	<<	B, I	5 << 3 = 40
Right-shift	>>	B, I	0x50 >> 0x03 = 0x0a
Is Range	is		A is B. The A operand is a Byte, and B is a Range symbol. The return value is a weight value from 0x00 (0.0) to 0xff (1.0).
Is Rule	is		A is B. The A operand is a Byte, and B is a Rule symbol. The return value is a fuzzy value which represents the accumulated result of processing A through each range in B.
Grouping	( )		2*(5+3) = 16
Array Access	[ ]		code[4] = the fifth value in the zero-based code array (zero, one, two, three, four)

For-next – see "While"

Function – see "Call"

If

```
<decide> := <if> [ <else> | <elseif> ]
<if> := 'if' '(' <expression> ')' '{' <block> '}'
<else> := 'else' '{' <block> '}'
<elseif> := 'elseif' '(' <expression> ')' '{' <block> '}'
               [ <else> | <elseif> ]


if (A > 10)
{ // Then branch... }
if (A > 20)
{ // Then branch... }
else
{ // Else branch... }

if (code == 1)
{}
elseif (code == 2)
{}
elseif (code == 3)
{}
else
{

        // Code is anything but 1, 2, or 3


}
```

The decision making system can be somewhat involved. The process begins with the if command, at which time the if expression is evaluated. If that evaluation returns a TRUE (non-zero) value, the if block is executed and the decision is complete.

If the if expression evaluates to FALSE (zero) then several things may occur. If there are no other branches in the if statement then the statement has finished and execution continues with the next statement. If there is an else branch, then the else block is executed.

When an if expression fails and there is an elseif branch, then the elseif expression is evaluated in the same way as the if expression was evaluated, with the same range of results. The elseif chain can continue to an indefinite length.

A case (or switch) statement can be emulated with a chain of if/elseif commands, though it will lack any of the optimization normally associated with a case statement.

Interrupt – see "Do"

Is

```
<is> := <expression> 'is' ( <range> | <rule> )
```

The is operator evaluates the result of an expression with respect to a fuzzy range or rule.

Is <range>

The expression evaluates to a byte value, and this value is fuzzified with respect to the named range. The returned activation value is a byte (0x00 to 0xff) which is the degree to which the value falls within the range. This can be used as a system weight to effect fuzzy assignments.

For example:

```
weight( speed is slow );
fuzval = 2@255;
```

This code can be read as, "to the extent that value 'speed' falls in the range 'slow,' fuzval takes the value of 2 (weight 1)". If speed is entirely outside of the range slow, fuzval is unaffected. If speed is entirely inside the range, fuzval accumulates the value 2@255 at face value. If speed is, say, 50% activated by the slow range, fuzval accumulates 2@255 at half its specified weight.

Is <rule>

The most common use of "is <range>" is to check a value against a set of ranges, to get a weighted average of their results. As a contrived example:

```
fuzvel as fuzzy;
velocity as real;

weight (speed is slow);
fuzvel = 1;
weight (speed is medium);
fuzvel = 3;
weight( speed is fast);
```

```
fuzvel = 8;
weight( speed is full);
fuzvel = 10;
velocity = fuzvel;
```

This code checks speed against a set of ranges (which presumably cover the entire possible span of speed values). It accumulates the weighted results in the fuzzy variable fuzvel. Once finished, fuzvel is evaluated to the crisp value velocity.

Fuzzy rules condense this process of testing and assignment down to two statements:

```
fuzvel as fuzzy;
velocity as real;

fuzvel = speed is speedrule;
velocity = fuzvel;
```

Or one statement, using the built-in type conversions:

```
velocity as real;
velocity = speed is speedrule;
```

See also "range" and "rule".

Loop – see "While"

Method

```
<method> := ( 'start'  | <identifier> ':' <identifier> )
       '(' [<prototype> [',' <prototype>]* ] ')'
       '{' <block> '}'

<prototype> := <identifier> 'as' <type>
```

A Fuzbol method consists of the qualified method name, a set of zero or more prototypes, and executable code. Note that the method definition mirrors the prototype list from its definition, but not the return type. Prototype lists are used to discriminate between different methods, but return types are not.

Fuzbol Methods may be functions, procedures, or interrupts. A function has a return value and may be used within expressions. Procedures do not return a value, but perform actions necessary to the program. Interrupts are procedures that are triggered by hardware events and lie outside the normal flow of control. Interrupt handlers should not be called as procedures, since they have a different return protocol.

A Fuzbol program begins execution at the special method named "start()". Start() may then call other methods as needed. Other methods must be defined as part of an object (as described under the "define" command), and may have zero or more parameters.

The body of the method is simply a block of statements.

See also "Call".

Method Prototype – see "Do"

Object – see "Define"

Out

```
<out> := ( ('outch' <expression>) | ('out' <expression>) | ('out' <string>) ) ';'

outch 'A';
outch 0x13;
out result;
out "string";
```

Fuzbol has three output commands. The first, "outch <expression>" evaluates an expression to the ASCII code for a single character. The second, "out <expression>" sends the result of the expression as an ASCII representation appropriate to the result type. The third, "out <string>" echoes a literal string to the UART.

When Fuzbol first initializes the MCU it sets the UART for serial communication at 9600 baud, 8-bits, 1 stop bit, no parity (on an 8MHz AT90S8515). With the addition of an RS232 level conversion chip, this link to the outside world can be piped to a serial LCD display or to the serial port on a desktop computer.

outch

Out-character casts the expression to a byte and then sends that byte, raw, to the UART. Suitable for printing single characters or special control codes.

out <byte>

Byte results are printed as an ASCII representation of the unsigned hexadecimal value (e.g. "0x3f").

out <integer>

Integer results are printed as an ASCII representation of the signed decimal value (e.g. "-1024").

out <real>

Real results are represented internally in a special floating-point format, and an ASCII representation of this raw data is sent to the UART. The hexadecimal mantissa is printed first, and this always starts with "1." The remaining mantissa values follow. Then, the specific power of 2 exponent is printed, prefixed by "e".

For example, the value 4.5 is displayed as "1.2000e02". The "e02" exponent indicates the mantissa is multiplied by $2^2 = 4$, for a hexadecimal value of 4.80. This is 4 plus 128/256ths which is 4.5 decimal.

out <fuzzy>

Fuzzy results are held in a similar binary fashion as reals. Out sends the value portion, an '@' separator, and the weight portion.

out <string>

String literals are useful as labels and other descriptive text. At this time, there is no provision for escape characters to allow embedding quotes or special characters in the string. These must be sent separately using outch.

Outch – see "Out"

Procedure – see "Call"

Prototype – see "Do"

Range

```
<range> := 'range' <identifier>
        '(' <expression> ',' <expression> ',' <expression> ')' ';'

range slow( 0x08, 0x02, 0x10 );
```

A range is defined by a unique identifier, plus three byte expressions. These three bytes define the range trapezoid by its center, plateau radius, and extent radius as shown in *Figure 13-1*. Ranges are defined outside of any method or object, and are global to a Fuzbol program.

Fuzzy ranges only encode byte values, from 0x00 to 0xff. The center places the triangle or trapezoid in this range, and any input that falls on the center is guaranteed to have a full activation weight. The plateau radius may be zero or some positive value, and it specifies the distance to either side of center a value may fall and still be considered fully activated. The extent radius, or width, must be at least as large as the plateau radius (and non-zero), and it specifies the width of the range to either side of center. A fuzzy membership diagram is shown in *Figure A-1.*

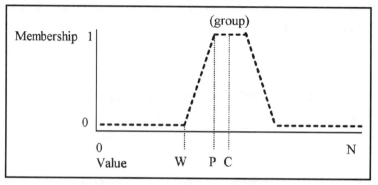

Figure A-1. Fuzzy Membership diagram.

Values are evaluated against a range with the "is" operator.

Rule

```
<rule> := 'rule' <identifier> '{' ( <identifier> 'as' <expression> ';' )* '}'

rule speedrule
{
slow as 0x00;
medium as 0x10;
fast as 0x40;
full as 0xFF;
}
```

A rule is identified by a unique name, and contains one or more range/result pairs. Ranges are defined outside of any method or object, and are global to a Fuzbol program.

The range body contains one or more lines, each of which references an existing range and the result value associated with that range.

Values are evaluated against each rule with the "is" operator.

Start() – see "Method"

Then – see "If"

Type – see "As"

Use

```
<use> := 'use' <filepath> ';'

use "uart.fuz";
use "subdir\filename.fuz";
```

The use command takes a string as its parameter.

Use forces Fuzbol to continue parsing with the first line of the specified file. After the last line of the nested file is parsed, parsing is continued where it left off in the parent file.

Use can be nested to any depth, though once a file has been referenced once any future references will be ignored so it is impossible to create file reference loops.

Variable – see "As"

Weight

```
<weight> := 'weight' '(' <expression> ')'
weight( 255 );
weight( value is really_low );
```

The system weight affects the weight of any fuzzy assignment. The system weight is multiplied against the weight of the right-hand-side fuzzy value in a fuzzy assignment.

System weights are preserved on the stack before a method is called, and restored after it returns. The called function or subroutine inherits the current system weight.

The weight() command simply replaces the previous system weight with the new value.

The weight expression evaluates to a byte, with a value from 0x00 (no weight) to 0xff (a unit weight of 1.0). A value of 0x7f is half of full weight.

While

```
<loop> := 'while' '(' <expression> ')' '{' <block> '}'

while (idx >= 0)
{
        // do stuff, hopefully changing idx in the process
}
```

A while loop is defined in two parts, the expression and a block of statements. When the while command is first encountered, the expression is evaluated. If this evaluation results in a TRUE (non-zero) result the block is executed. This cycle is repeated until such time as the expression evaluates to FALSE (zero).

A "for-next" loop can be emulated as follows:

```
index = start;
while (index <= end)
{
        // various other statements...

index = index + step;
}
```

Appendix B: Sample Programs

Alive.asm

```
;=====================================================================
;
;       Alive!
;
;       NOTE:   This program uses Port-C, which is not available
;                       on the SRS 8515 companion board.
;
;=====================================================================

.include "8515def.inc"

;=====================================================================
;
; Globals
;
.DEF Temp                   = R16

.EQU PORTC_HEART       = 0                  ; Heartbeat flag Port C Bit 0

;=====================================================================

.CSEG
.ORG 0x00

        rjmp    RESET                       ; Power-on Reset
        rjmp    RESET                       ; IRQ0
        rjmp    RESET                       ; IRQ1
        rjmp    RESET                       ; Timer1 Capture
        rjmp    RESET                       ; Timer1 Compare A
        rjmp    RESET                       ; Timer1 Compare B
        rjmp    RESET                       ; Timer1 Overflow
        rjmp    RESET                       ; Timer0 Overflow
        rjmp    RESET                       ; SPI Transfer Complete
        rjmp    RESET                       ; UART RX Complete
        rjmp    RESET                       ; UDR Empty
        rjmp    RESET                       ; UART TX Complete
        rjmp    RESET                       ; Analog Comparator
```

```
;=======================================================================

.ORG 0x0d

RESET:
        ; -------------------------------
        ; Initialize the Stack Pointer!
        ;
        ldi             Temp,   low(RAMEND)
        out             SPL,    Temp    ;init Stack Pointer
        ldi             Temp,   high(RAMEND)
        out             SPH,    Temp

        ; -------------------------------
        ; Indicator light on Port C

        ;
        sbi             DDRC, PORTC_HEART               ; Set Data Direction to OUTPUT
        sbi             PORTC, PORTC_HEART              ; Turn on the LED
        rjmp            MainLoop                        ; Go to the main program

;=======================================================================
; Main Loop; does nothing

;

MainLoop:
        rjmp    MainLoop
```

mcu_heart.asm

```
;=======================================================================
;
;       Robotic Heartbeat
;
;       Demonstrates:
;               Timer
;               Interrupts
;               Digital output
;
;       NOTE:   This program uses Port-C, which is not available
;                       on the SRS 8515 companion board.
;=======================================================================

.include "8515def.inc"

;=======================================================================
;
; Put these into the upper-bank registers, so we can LDI them
;
.DEF Temp               = R16
.DEF BeatNum            = R17
;=======================================================================
.EQU PORTC_HEART        = 0     ; Heartbeat flag Port C Bit 0
.EQU BEAT_PER_INT       = 24    ; Roughly 5 beats per second
;=======================================================================
.CSEG
.ORG 0x00
```

```
        rjmp    RESET                   ; Power-on Reset
        rjmp    RESET                   ; IRQ0
        rjmp    RESET                   ; IRQ1
        rjmp    RESET                   ; Timer1 Capture
        rjmp    RESET                   ; Timer1 Compare A
        rjmp    RESET                   ; Timer1 Compare B
        rjmp    RESET                   ; Timer1 Overflow
        rjmp    Heartbeat               ; Timer0 Overflow
        rjmp    RESET                   ; SPI Transfer Complete
        rjmp    RESET                   ; UART RX Complete
        rjmp    RESET                   ; UDR Empty
        rjmp    RESET                   ; UART TX Complete
        rjmp    RESET                   ; Analog Comparator


;=====================================================================

.ORG 0x0d

RESET:
        ; -----------------------------------
        ; Initialize the Stack Pointer!
        ;
        ldi             Temp,   low(RAMEND)
        out             SPL,    Temp    ;init Stack Pointer
        ldi             Temp,   high(RAMEND)
        out             SPH,    Temp


        ; -----------------------------------
        ; Indicator light on Port C
        ;
        sbi             DDRC, PORTC_HEART           ; Set Data Direction to OUTPUT
        sbi             PORTC, PORTC_HEART          ; Start with Heartbeat ON
        ; -----------------------------------
        ; Setup the heartbeat timer
        ;
        ldi             Temp, 4
        out             TCCR0, Temp                 ; Pre-scale Timer0 by 256

        ldi             Temp, 2                     ; TOIE0=1 (timer 0 interrupt on)
        out             TIMSK, Temp

        ldi             BeatNum, BEAT_PER_INT


        ; ---------------------------------------------
        sei                                         ; Activate global interrupts
        rjmp    MainLoop                            ; Go to the main program

;=====================================================================
; Main Loop; does nothing
;
MainLoop:
        rjmp    MainLoop
;=====================================================================
;       Heartbeat
;
; Toggles the state of the heartbeat indicator
```

```
;
; This interrupt is called each time the heartbeat timer overflows.
; The system clock is 8,000,000Hz, with a pre-scale of 256 which
; gives 31,250 timer ticks per second.  The 8-bit counter will then
; overflow every 256 ticks, or about 122 times a second.
;
; We then manage this overflow with our own counter, giving us
; 5 pulses per second, or a pulse once every 24 interrupts.
;
Heartbeat:
        dec             BeatNum
        brne            hb_done                 ; If not zero, early exit

        ldi             BeatNum, BEAT_PER_INT ; Reset counter

        ; Toggle heartbeat by sensing current state and reversing it
        ;
        sbis    PORTC, PORTC_HEART              ; Skip next if heartbeat is set...
        rjmp    hb_set                          ; clear, so set it...
hb_clear:
        cbi     PORTC, PORTC_HEART      ; Clear the heartbeat
        rjmp    hb_done
hb_set:
        sbi     PORTC, PORTC_HEART              ; Set the heartbeat
hb_done:
        reti                                    ; Return from interrupt
```

mcu_feeler.asm

```
;=======================================================================
;
;       Feeler Switches
;
;       Heartbeat plus...
;
;       Demonstrates:
;               Digital Input with internal pull-up
;               Additional use of interrupts
;
;       NOTE:   This program uses Port-C, which is not available
;                       on the SRS 8515 companion board.
;=======================================================================

.include "8515def.inc"

;=======================================================================
;
; Put these into the upper-bank registers, so we can LDI them
;
.DEF Temp           = R16
.DEF BeatNum    = R17


;=======================================================================

.EQU PORTC_HEART        = 0             ; Heartbeat flag Port C Bit 0
.EQU PORTC_LFI          = 1             ; Left Feeler Input
.EQU PORTC_RFI          = 2             ; Right Feeler Input
.EQU PORTC_LFO          = 3             ; Left Feeler Output
```

```
.EQU PORTC_RFO        = 4                  ; Right Feeler Output

.EQU BEAT_PER_INT     = 24    ; Roughly 5 beats per second

;===================================================================

.CSEG
.ORG 0x00

        rjmp    RESET                   ; Power-on Reset
        rjmp    RESET                   ; IRQ0
        rjmp    RESET                   ; IRQ1
        rjmp    RESET                   ; Timer1 Capture
        rjmp    RESET                   ; Timer1 Compare A
        rjmp    RESET                   ; Timer1 Compare B
        rjmp    RESET                   ; Timer1 Overflow
        rjmp    Heartbeat               ; Timer0 Overflow
        rjmp    RESET                   ; SPI Transfer Complete
        rjmp    RESET                   ; UART RX Complete
        rjmp    RESET                   ; UDR Empty
        rjmp    RESET                   ; UART TX Complete
        rjmp    RESET                   ; Analog Comparator

;===================================================================

.ORG 0x0d

RESET:
        ; ---------------------------------
        ; Initialize the Stack Pointer!
        ;
        ldi         Temp,   low(RAMEND)
        out         SPL,    Temp                 ;init Stack Pointer
        ldi         Temp,   high(RAMEND)
        out         SPH,    Temp
        ; ---------------------------------
        ; Heartbeat light on Port C
        ;
        sbi         DDRC, PORTC_HEART        ; Set Data Direction to OUTPUT
        sbi         PORTC, PORTC_HEART       ; Start with Heartbeat ON
        ; ---------------------------------
        ; Feeler inputs on Port C
        ;
        ;Both ports set to Inputs, with internal pull-up
        ;
        cbi         DDRC, PORTC_LFI
        sbi         PORTC, PORTC_LFI

        cbi         DDRC, PORTC_RFI
        sbi         PORTC, PORTC_RFI


        ; --------------------------------
        ; Feeler lights on Port C
        ;
        ;Both ports set to output, with the light off
        ;
        sbi         DDRC, PORTC_LFO
        cbi         PORTC, PORTC_LFO
```

```
        sbi             DDRC, PORTC_RFO
        cbi             PORTC, PORTC_RFO


        ; Note that a short Port-C initialization routine is given below.
        ; This code takes less than half the instructions, yet preserves
        ; the flexibility of the symbolic constants for the port bits.
        ;
        ;       ldi     Temp, (1<<PORTC_HEART) + (1<<PORTC_LFO) + (1<<PORTC_RFO)
        ;       out     DDRC, Temp
        ;
        ;       ldi Temp, (1<<PORTC_HEART) + (1<<PORTC_LFI) + (1<<PORTC_RFI)
        ;       out PORTC, Temp


        ; ---------------------------------
        ; Setup the heartbeat timer
        ;

        ldi             Temp, 4
        out             TCCR0, Temp                 ; Pre-scale Timer0 by 256

        ldi             Temp, 2                     ; TOIE0=1 (timer 0 interrupt on)
        out             TIMSK, Temp

        ldi             BeatNum, BEAT_PER_INT


        ; ---------------------------------
        sei                                         ; Activate global interrupts
        rjmp    MainLoop                            ; Go to the main program

;====================================================================
; Main Loop
;
;       Polls the feelers, and turns on the lights if they are touched
;
MainLoop:
        sbis    PINC, PORTC_LFI             ; Skip of left feeler is set (not touched)
        sbi             PORTC, PORTC_LFO    ; touched, so set the light

        sbis    PINC, PORTC_RFI             ; Right feeler touched?
        sbi             PORTC, PORTC_RFO    ; yes, set light


        rjmp    MainLoop


;====================================================================
;       Heartbeat
;
; Toggles the state of the heartbeat indicator
;
; This interrupt is called each time the heartbeat timer overflows.
; The system clock is 8,000,000Hz, with a pre-scale of 256 which
; gives 31,250 timer ticks per second.  The 8-bit counter will then
; overflow every 256 ticks, or about 122 times a second.
;
; We then manage this overflow with our own counter, giving us
; 5 pulses per second, or a pulse once every 24 interrupts.
;
Heartbeat:
        dec             BeatNum
        brne            hb_done                     ; If not zero, early exit
```

```
        ldi             BeatNum, BEAT_PER_INT ; Reset counter

        ; Toggle heartbeat by sensing current state and reversing it
        ;
        sbis    PORTC, PORTC_HEART              ; Skip next if heartbeat is set...
        rjmp    hb_set                          ; clear, so set it...

hb_clear:
        cbi     PORTC, PORTC_HEART              ; Clear the heartbeat

        sbic    PINC, PORTC_LFI                 ; Skip of left feeler is clear (touched)
        cbi     PORTC, PORTC_LFO                ; Clear left feeler output

        sbic    PINC, PORTC_RFI                 ; Right feeler free?
        cbi     PORTC, PORTC_RFO                ; Clear right feeler output

        rjmp    hb_done

hb_set:
        sbi     PORTC, PORTC_HEART              ; Set the heartbeat

hb_done:
        reti                                    ; Return from interrupt
```

mcu_motor.asm

```
;=====================================================================
;
;       PWM Motor Control
;
;       Heartbeat plus
;       Feelers plus
;
;       Demonstrates:
;               Dual PWM output
;               Simple state-machine logic
;
;       NOTE:   This program uses Port-C, which is not available
;               on the SRS 8515 companion board.
;
;=====================================================================

.include "8515def.inc"

;=====================================================================
;
.DEF Temp            = R16          ; Generic temporary variable
.DEF BeatNum    = R17          ; Heartbeat interrupt counter
.DEF FeelerOn = R18             ; Handy reference for feeler status

;=====================================================================

.EQU PORTC_HEART        = 0            ; Heartbeat flag Port C Bit 0
.EQU PORTC_LFI          = 1            ; Left Feeler Input
.EQU PORTC_RFI          = 2            ; Right Feeler Input
.EQU PORTC_LFO          = 3            ; Left Feeler Output
```

```
.EQU PORTC_RFO          = 4             ; Right Feeler Output
.EQU PORTC_LREV         = 5             ; Left motor Reverse
.EQU PORTC_RREV         = 6             ; Right motor Reverse

.EQU BEAT_PER_INT       = 61            ; Changed to roughly 2 beats per second

.EQU LEFT_MOTOR         = OCR1AL        ; First PWM output port
.EQU RIGHT_MOTOR        = OCR1BL        ; Second PWM output port

.EQU MOTOR_OFF          = 0             ; Motor speeds, handy reference
.EQU MOTOR_MED          = 96
.EQU MOTOR_ON           = 255

.EQU TOUCH_NONE         = 0             ; Flag definitions for FeelerOn
.EQU TOUCH_LEFT         = 1
.EQU TOUCH_RIGHT        = 2
.EQU TOUCH_BOTH         = TOUCH_LEFT | TOUCH_RIGHT


;====================================================================

.CSEG
.ORG 0x00

        rjmp    RESET                   ; Power-on Reset
        rjmp    RESET                   ; IRQ0
        rjmp    RESET                   ; IRQ1
        rjmp    RESET                   ; Timer1 Capture
        rjmp    RESET                   ; Timer1 Compare A
        rjmp    RESET                   ; Timer1 Compare B
        rjmp    RESET                   ; Timer1 Overflow
        rjmp    Heartbeat               ; Timer0 Overflow
        rjmp    RESET                   ; SPI Transfer Complete
        rjmp    RESET                   ; UART RX Complete
        rjmp    RESET                   ; UDR Empty
        rjmp    RESET                   ; UART TX Complete
        rjmp    RESET                   ; Analog Comparator


;====================================================================

.ORG 0x0d

RESET:
        ; ---------------------------
        ; Initialize the Stack Pointer!
        ;
        ldi             Temp,   low(RAMEND)
        out             SPL,    Temp                    ;init Stack Pointer
        ldi             Temp,   high(RAMEND)
        out             SPH,    Temp

        ; ---------------------------
        ;       This short-form Port-C initialization sets the lights
        ;       and motor direction controls to outputs and the sensors
        ;       to inputs.
        ;
        ldi     Temp, (1<<PORTC_HEART) + (1<<PORTC_LFO) + (1<<PORTC_RFO) + (1<<PORTC_LREV)
+ (1<<PORTC_RREV)
        out     DDRC, Temp
```

```
        ldi Temp, (1<<PORTC_HEART) + (1<<PORTC_LFI) + (1<<PORTC_RFI)
        out PORTC, Temp

        ; ---------------------------
        ; Setup the heartbeat timer
        ;
        ldi             Temp, 4
        out             TCCR0, Temp             ; Pre-scale Timer0 by 256; 31KHz clock

        ldi             Temp, 2
        out             TIMSK, Temp             ; TOIE0=1 (timer 0 interrupt on)

        ldi             BeatNum, BEAT_PER_INT

        ; -------------------------------
        ; Setup the dual PWM timers
        ;
        sbi             DDRD, 5                 ; Force the PD5 to be output, no pullup,
        cbi             PORTD, 5                ; to enable it as a PWM port

        ldi             Temp, 0xa1
        out             TCCR1A, Temp            ; Set PWM mode on both counters

        ldi             Temp, 0x02
        out             TCCR1B, Temp            ; Pre-scale by 8 on 8MHz clock

        ldi             Temp, 0x00
        out             OCR1AH, Temp            ; Clear the high-bytes of the compare
        out             OCR1BH, Temp            ; register; only using 8-bit PWM

        ldi             Temp, MOTOR_OFF         ; Start with both motors off...
        out             LEFT_MOTOR, Temp
        out             RIGHT_MOTOR, Temp

        ; ---------------------------------
        sei                                     ; Activate global interrupts
        rjmp    MainLoop                        ; Go to the main program

;======================================================================
; Main Loop
;
; No loops, but starts the state machine...
;
MainLoop:
        ldi             FeelerOn, 0x00          ; Initialize this to zero
        rcall           Feelers                 ; get initial state of the feelers

; ----------- Go forward, enter the forward state
forward:
        cbi             PORTC, PORTC_LREV       ; Signal the motors to go forwards
        cbi             PORTC, PORTC_RREV

        ldi             Temp, MOTOR_MED
        out             LEFT_MOTOR, Temp
        out             RIGHT_MOTOR, Temp
```

```
state_fw:
      cpi             FeelerOn, TOUCH_RIGHT
      breq            turn_left
      cpi             FeelerOn, TOUCH_LEFT
      breq    turn_right
      cpi             FeelerOn, TOUCH_BOTH
      breq            stop_reverse
      ; TOUCH_NONE
      rcall   Feelers
      rjmp    state_fw

; ---------- Turn left, enter the left-turning state
turn_left:
      ldi             Temp, MOTOR_MED
      out             RIGHT_MOTOR, Temp
      ldi             Temp, MOTOR_OFF
      out             LEFT_MOTOR, Temp
state_tl:
      cpi             FeelerOn, TOUCH_NONE
      breq    forward
      cpi             FeelerOn, TOUCH_BOTH
      breq    stop_reverse
      ; TOUCH_RIGHT
      rcall   Feelers
      rjmp    state_tl

; --------- Turn right, enter the right-turning state
turn_right:
      ldi             Temp, MOTOR_OFF
      out             RIGHT_MOTOR, Temp
      ldi             Temp, MOTOR_MED
      out             LEFT_MOTOR, Temp
state_tr:
      cpi             FeelerOn, TOUCH_NONE
      breq    forward
      cpi             FeelerOn, TOUCH_BOTH
      breq    stop_reverse
      ; TOUCH_LEFT
      rcall   Feelers
      rjmp    state_tr

; --------- Stop prior to reversing, enter the stop/reverse state
stop_reverse:
      ldi             Temp, MOTOR_OFF
      out             RIGHT_MOTOR, Temp
      out             LEFT_MOTOR, Temp
state_sr1:
      sbic    PORTC, PORTC_HEART
      rjmp    state_sr1
state_sr2:
      sbis    PORTC, PORTC_HEART
      rjmp    state_sr2
state_sr3:
      sbic    PORTC, PORTC_HEART
      rjmp    state_sr3

      rjmp    reverse

; ---------Go backwards, enter the reverse state
reverse:
```

```
        sbi             PORTC, PORTC_LREV           ; Signal the motors to reverse
        sbi             PORTC, PORTC_RREV

        ldi             Temp, MOTOR_MED
        out             LEFT_MOTOR, Temp
        out             RIGHT_MOTOR, Temp
state_rv:
        cpi             FeelerOn, TOUCH_NONE
        breq    stop_forward
        rcall   Feelers
        rjmp    state_rv

; --------Stop prior to forward motion, enter the stop/forward state
stop_forward:
        ldi             Temp, MOTOR_OFF
        out             RIGHT_MOTOR, Temp
        out             LEFT_MOTOR, Temp
state_sf1:
        sbic    PORTC, PORTC_HEART
        rjmp    state_sf1
state_sf2:
        sbis    PORTC, PORTC_HEART
        rjmp    state_sf2
state_sf3:
        sbic    PORTC, PORTC_HEART
        rjmp    state_sf3

        rjmp    forward


;=================================================================
;       Feelers
;
; Poll the feelers and update the status port and internal flag
;
Feelers:
        sbic    PINC, PORTC_LFI        ; Touched?
        rjmp    feel_right             ; No, check right
        sbi     PORTC, PORTC_LFO       ; Yes, so set the light
        sbr     FeelerOn, TOUCH_LEFT

feel_right:
        sbic    PINC, PORTC_RFI        ; Touched?
        rjmp    feel_done              ; No, done
        sbi     PORTC, PORTC_RFO       ; Yes, set light
        sbr     FeelerOn, TOUCH_RIGHT

feel_done:
        ret


;=================================================================
;       Heartbeat
;
; Toggles the state of the heartbeat indicator
;
; This interrupt is called each time the heartbeat timer overflows.
; The system clock is 8,000,000Hz, with a pre-scale of 256 which
; gives 31,250 timer ticks per second.  The 8-bit counter will then
; overflow every 256 ticks, or about 122 times a second.
;
; We then manage this overflow with our own counter, BeatNum,
```

```
; giving one heartbeat every BEAT_PER_INT interrupts.
;
Heartbeat:
        dec             BeatNum
        brne            hb_done                 ; If not zero, early exit

        ldi             BeatNum, BEAT_PER_INT ; Reset counter

        ; Toggle heartbeat by sensing current state and reversing it
        ;
        sbis    PORTC, PORTC_HEART              ; Skip next if heartbeat is set...
        rjmp    hb_set                          ; clear, so set it...

        ; ------------------
        ; Clear the heartbeat
        ;
hb_clear:
        cbi     PORTC, PORTC_HEART              ; Clear the heartbeat

        ;
        ; As we clear the heartbeat, see if we can clear the feelers too...
        ;
        sbis    PINC, PORTC_LFI         ; Touched?
        rjmp    hb_fr                   ; Yes, check right
        cbi     PORTC, PORTC_LFO        ; No, clear left feeler output
        cbr     FeelerOn, TOUCH_LEFT

hb_fr:
        sbis    PINC, PORTC_RFI         ; Touched?
        rjmp    hb_done                 ; Yes, done
        cbi     PORTC, PORTC_RFO        ; No, clear right feeler output
        cbr     FeelerOn, TOUCH_RIGHT
        rjmp    hb_done

        ; -----------
        ; Set the heartbeat
        ;
hb_set:
        sbi     PORTC, PORTC_HEART              ; Set the heartbeat

hb_done:
        reti                                    ; Return from interrupt
```

mcu_eyes.asm

```
;=====================================================================
;
;       Eyes
;
;       Demonstrates:
;               Software serial interface ("bit banging")
;
;       NOTE:  This program uses Port-C, which is not available
;                       on the SRS 8515 companion board.
;=====================================================================

.include "8515def.inc"


;=====================================================================
```

```
;
; Globals
;
.DEF LeftSpeed          = R0
.DEF RightSpeed         = R1
.DEF Temp               = R16
.DEF BeatNum            = R17
.DEF FeelerOn           = R18    ; Handy reference for feeler status


;
; Local to Analog-to-Digital conversion
;
.DEF adData             = R19   ; Channel to convert, then data from that channel
.DEF ad_delay           = R20   ; Temporary delay counter
.DEF ad_bit             = R21   ; Temporary bit counter


;====================================================================

.EQU PORTC_HEART        = 0      ; Heartbeat flag Port C Bit 0

.EQU PORTC_LFI          = 1      ; Left Feeler Input
.EQU PORTC_RFI          = 2      ; Right Feeler Input
.EQU PORTC_LFO          = 3      ; Left Feeler Output
.EQU PORTC_RFO          = 4      ; Right Feeler Output
.EQU PORTC_LREV         = 5      ; Left motor Reverse
.EQU PORTC_RREV         = 6      ; Right motor Reverse

; Analog-to-Digital Port B bits
.EQU PORTB_SELECT       = 0
.EQU PORTB_MOSI         = 1
.EQU PORTB_MISO         = 2
.EQU PORTB_CLK          = 3


.EQU BEAT_PER_INT       = 61

.EQU spiBuffer          = 0x60              ; Start buffer at first decent SRAM location

.EQU LEFT_MOTOR         = OCR1AL            ; First PWM output port
.EQU RIGHT_MOTOR        = OCR1BL            ; Second PWM output port

.EQU MOTOR_OFF          = 0                 ; Motor speeds, handy reference
.EQU MOTOR_MED          = 96
.EQU MOTOR_ON           = 255

.EQU TOUCH_NONE         = 0                 ; Flag definitions for FeelerOn
.EQU TOUCH_LEFT         = 1
.EQU TOUCH_RIGHT        = 2
.EQU TOUCH_BOTH         = TOUCH_LEFT | TOUCH_RIGHT


;====================================================================

.CSEG
.ORG 0x00

        rjmp    RESET                   ; Power-on Reset
        rjmp    RESET                   ; IRQ0
        rjmp    RESET                   ; IRQ1
        rjmp    RESET                   ; Timer1 Capture
        rjmp    RESET                   ; Timer1 Compare A
```

```
        rjmp    RESET                   ; Timer1 Compare B
        rjmp    RESET                   ; Timer1 Overflow
        rjmp    Heartbeat               ; Timer0 Overflow
        rjmp    RESET                   ; SPI Transfer Complete
        rjmp    RESET                   ; UART RX Complete
        rjmp    RESET                   ; UDR Empty
        rjmp    RESET                   ; UART TX Complete
        rjmp    RESET                   ; Analog Comparator
```

;==

```
.ORG 0x0d

RESET:
        ; ------------------------------
        ; Initialize the Stack Pointer!
        ;
        ldi             Temp,   low(RAMEND)
        out             SPL,    Temp                    ;init Stack Pointer
        ldi             Temp,   high(RAMEND)
        out             SPH,    Temp


        ; ----------------------------
        ;       This short-form Port-C initialization sets the lights
        ;       and motor direction controls to outputs and the sensors
        ;       to inputs.
        ;
        ldi     Temp, (1<<PORTC_HEART) + (1<<PORTC_LFO) + (1<<PORTC_RFO) + (1<<PORTC_LREV)
+ (1<<PORTC_RREV)
        out     DDRC, Temp

        ldi Temp, (1<<PORTC_HEART) + (1<<PORTC_LFI) + (1<<PORTC_RFI)
        out PORTC, Temp


        ; ------------------
        ; Heartbeat timer
        ;
        ldi             Temp, 4
        out             TCCR0, Temp             ; Pre-scale Timer0 by 256

        ldi             Temp, 2                 ; TOIE0=1 (timer 0 interrupt on)
        out             TIMSK, Temp

        ldi             BeatNum, BEAT_PER_INT

        ;---------------------------
        ; Setup the dual PWM timers
        ;
        sbi             DDRD, 5                 ; Force the PD5 to be output, no pullup,
        cbi             PORTD, 5                ; to enable it as a PWM port

        ldi             Temp, 0xa1
        out             TCCR1A, Temp            ; Set PWM mode on both counters

        ldi             Temp, 0x02
        out             TCCR1B, Temp            ; Pre-scale by 8 on 8MHz clock

        ldi             Temp, 0x00
        out             OCR1AH, Temp            ; Clear the high-bytes of the compare
        out             OCR1BH, Temp            ; register; only using 8-bit PWM
```

```
        ldi             Temp, MOTOR_OFF        ; Start with both motors off...
        out             LEFT_MOTOR, Temp
        out             RIGHT_MOTOR, Temp

ldi    Temp, 0xff
out DDRA, Temp
out PORTA, Temp

        ; --------------------
        ; Init AtoD
        ;
        ldi             Temp, 0b1011
        out             DDRB, Temp             ; PB0,1,3 out; PB2 in
        clr             Temp
        out             PORTB, Temp
        sbi             PORTB, PORTB_SELECT

        ; ---------------------
        sei                                    ; Activate global interrupts
        rjmp    MainLoop                       ; Go to the main program

;===================================================================
; Main Loop, does nothing
;
MainLoop:
        ldi             Temp, MOTOR_MED
        mov             LeftSpeed, Temp
        mov             RightSpeed, Temp

        ldi             FeelerOn, 0x00         ; Initialize this to zero
        rcall   Feelers                        ; get initial state of the feelers

; --------Go forward, enter the forward state
forward:
        cbi             PORTC, PORTC_LREV      ; Signal the motors to go forwards
        cbi             PORTC, PORTC_RREV

state_fw:
        out     LEFT_MOTOR, LeftSpeed
        out     RIGHT_MOTOR, RightSpeed

        cpi     FeelerOn, TOUCH_RIGHT
        breq    turn_left
        cpi     FeelerOn, TOUCH_LEFT
        breq    turn_right
        cpi     FeelerOn, TOUCH_BOTH
        breq    stop_reverse
        ; TOUCH_NONE
        rcall   Feelers
        rjmp    state_fw

; ------- Turn left, enter the left-turning state
turn_left:
        ldi     Temp, MOTOR_MED
        out     RIGHT_MOTOR, Temp
        ldi     Temp, MOTOR_OFF
        out     LEFT_MOTOR, Temp
state_tl:
        cpi     FeelerOn, TOUCH_NONE
```

```
        breq    forward
        cpi     FeelerOn, TOUCH_BOTH
        breq    stop_reverse
        ; TOUCH_RIGHT
        rcall   Feelers
        rjmp    state_tl

; ——— Turn right, enter the right-turning state
turn_right:
        ldi     Temp, MOTOR_OFF
        out     RIGHT_MOTOR, Temp
        ldi     Temp, MOTOR_MED
        out     LEFT_MOTOR, Temp
state_tr:
        cpi     FeelerOn, TOUCH_NONE
        breq    forward
        cpi     FeelerOn, TOUCH_BOTH
        breq    stop_reverse
        ; TOUCH_LEFT
        rcall   Feelers
        rjmp    state_tr

; ——— Stop prior to reversing, enter the stop/reverse state
stop_reverse:
        ldi     Temp, MOTOR_OFF
        out     RIGHT_MOTOR, Temp
        out     LEFT_MOTOR, Temp
state_sr1:
        sbic    PORTC, PORTC_HEART
        rjmp    state_sr1
state_sr2:
        sbis    PORTC, PORTC_HEART
        rjmp    state_sr2
state_sr3:
        sbic    PORTC, PORTC_HEART
        rjmp    state_sr3

        rjmp    reverse

; ——— Go backwards, enter the reverse state
reverse:
        sbi     PORTC, PORTC_LREV               ; Signal the motors to reverse
        sbi     PORTC, PORTC_RREV

        ldi     Temp, MOTOR_MED
        out     LEFT_MOTOR, Temp
        out     RIGHT_MOTOR, Temp
state_rv:
        cpi     FeelerOn, TOUCH_NONE
        breq    stop_forward
        rcall   Feelers
        rjmp    state_rv

; ——— Stop prior to forward motion, enter the stop/forward state
stop_forward:
        ldi     Temp, MOTOR_OFF
        out     RIGHT_MOTOR, Temp
        out     LEFT_MOTOR, Temp
state_sf1:
        sbic    PORTC, PORTC_HEART
```

```
        rjmp    state_sf1
state_sf2:
        sbis    PORTC, PORTC_HEART
        rjmp    state_sf2
state_sf3:
        sbic    PORTC, PORTC_HEART
        rjmp    state_sf3

        rjmp    forward
```

```
;========================================================================
;       Feelers
;
; Poll the feelers and update the status port and internal flag
; Also, check the eyes and update the speed based on the light
; sources.
;
Feelers:
        ldi     adData, 0
        rcall   adConvert
        lsl     adData
        lsl     adData
        mov     RightSpeed, adData

        sbic    PINC, PORTC_LFI         ; Touched?
        rjmp    feel_right              ; No, check right
        sbi     PORTC, PORTC_LFO        ; Yes, so set the light
        sbr     FeelerOn, TOUCH_LEFT

feel_right:
        ldi     adData, 1
        rcall   adConvert
        lsl     adData
        lsl     adData
        mov     LeftSpeed, adData

        sbic    PINC, PORTC_RFI         ; Touched?
        rjmp    feel_done               ; No, done
        sbi     PORTC, PORTC_RFO        ; Yes, set light
        sbr     FeelerOn, TOUCH_RIGHT

feel_done:
        ret
```

```
;========================================================================
;       Heartbeat
;
; Toggles the state of the heartbeat indicator
;
; This interrupt is called each time the heartbeat timer overflows.
; The system clock is 8,000,000Hz, with a pre-scale of 256 which
; gives 31,250 timer ticks per second.  The 8-bit counter will then
; overflow every 256 ticks, or about 122 times a second.
;
; We then manage this overflow with our own counter, BeatNum,
; giving one heartbeat every BEAT_PER_INT interrupts.
;
Heartbeat:
        dec     BeatNum
        brne    hb_done                 ; If not zero, early exit
```

```
        ldi     BeatNum, BEAT_PER_INT ; Reset counter

        ; Toggle heartbeat by sensing current state and reversing it
        ;
        sbis    PORTC, PORTC_HEART      ; Skip next if heartbeat is set...
        rjmp    hb_set                  ; clear, so set it...
hb_clear:
        cbi     PORTC, PORTC_HEART      ; Clear the heartbeat
        rjmp    hb_done
hb_set:
        sbi     PORTC, PORTC_HEART      ; Set the heartbeat

hb_done:
        reti                            ; Return from interrupt


;=====================================================================
;       adConvert
;
; Analog-to-Digital conversion
;
; The calling code sets adData to the channel to convert (0..7).
; The converted value is in adData on exit
;
adConvert:
        cbi     PORTB, PORTB_CLK
        cbi     PORTB, PORTB_SELECT
        rcall   ad_wait

        ; --------------------
        ; Send the command out
        ;
ad_send:
        ori     adData, 0x18        ; Start, single ended
        ldi     ad_bit, 8           ; Send all 8 bits; only need last 5 though

ads_tst:
        lsl     adData
        brcc    ads_0
ads_1:
        sbi     PORTB, PORTB_MOSI
        rjmp    ads_clk
ads_0:
        cbi     PORTB, PORTB_MOSI

ads_clk:
        rcall   ad_clock

        dec     ad_bit
        brne    ads_tst

        ; ————
        ; An extra clock cycle
        ;
        rcall   ad_clock

        ; ————
        ; Receive the data
        ;
```

```
ad_receive:
        ldi     ad_bit, 8
        clr     adData

adr_tst:
        rcall   ad_clock

        lsl     adData
        sbic    PINB, PORTB_MISO
        sbr     adData, 0x01

        dec     ad_bit
        brne    adr_tst

ad_end:
        rcall   ad_wait
        sbi     PORTB, PORTB_SELECT
        rcall   ad_wait

        ret


;       —————
;       ad_wait
;
;       3 cycles on call
;       8 cycles in loop
;       4 cycles on return
;       = 15 cycle delay
;
;       at 8,000,000 cycles per second, gives
;       roughly 533,333 waits per second,
;       for about 260,000 conversions per second
;
ad_wait:
                                        ; 3 cycles to call
        ldi     ad_delay, 3             ; ... total 8 cycle loop
adw0:
        dec     ad_delay                ; 1 cycle
        brne    adw0                    ; 2 cycles branch, 1 cycle fall-through
        ret                             ; 4 cycles to return

ad_clock:
        rcall   ad_wait                 ; setup
        sbi     PORTB, PORTB_CLK
        rcall   ad_wait
        cbi     PORTB, PORTB_CLK

        ret
```

Heartbeat.fuz

```
// =======================================================================
//
//              Heartbeat.fuz
//
//      Interrupt Driven Heartbeat
//
//      Copyright (c) 1999 Edwin Wise and Simulated Reality Systems
//      http://www.simreal.com/Fuzbol
```

```
//
// =================================================================
use "system_8515.fuz";

// =================================================================

define Heart
{
        count as integer;                       // Count interrupts

        do Init();                              // Initialize Heartbeat timer

        do Pulse() as interrupt 7;              // Interrupt handler on Timer-0
                const PULSE_CNT = 5;            // How many interrupts per pulse
                const ON = 0x10;                // OR to turn on
}

Heart:Init()
{
        count = 0;

        System:DDRD = System:PIN4_OUT;          // Output
        System:PORTD = System:PORTD & ~ON;      // Off

        System:TCNT0 = 0;                       // Init counter to 0
        System:TCCR0 = System:T0_CK1024;        // Clock step once every 1024 cycles
        System:TIMSK = System:TOIE0;            // Enable Timer-0 overflow interrupt
}

Heart:Pulse()
{
        count = count - 1;
        if (count <= 0)
        {
                count = PULSE_CNT;
                if (System:PORTD & ON)
                {
                        System:PORTD = System:PORTD & ~ON;
                }
                else
                {
                        System:PORTD = System:PORTD | ON;
                }
        }
}

// =================================================================
//      start
//
//      The main program does little more than initialize interrupts
//      and die — all of the work is done by the interrupt handlers.
//
start()
{
        write as memory 0x2001;
        read as byte;

        outch 0x0d;
        out "(heartbeat)";
```

```
        outch 0x0d;

        Heart:Init();
}
```

Legs.fuz

```
// ==================================================================
//
//              Legs.fuz
//
//      Basic motor driver and feeler interrupts...
//
//      NOTE:   This program uses Port-C, which is not available
//                      on the SRS 8515 companion board.
//
//      Copyright (c) 1999 Edwin Wise and Simulated Reality Systems
//      http://www.simreal.com/Fuzbol
// ==================================================================

use "system_8515.fuz";

// ==================================================================

define Robot
{
        beat as byte;                   // Heartbeat counter
                const PULSE_CNT = 10;
                const HEART_ON = 0x10;

        left_vel as byte;               // Left motor velocity
        right_vel as byte;              // Right motor velocity
                const VEL_ADD = 8;
                const VEL_MAX = 128;

                const LEFT_REV = 0x80;
                const RIGHT_REV = 0x40;

        do Init();                      // Initialize Heartbeat timer
        do Pulse() as interrupt 7;      // Interrupt handler on Timer-0
        do LeftFeeler() as interrupt 1; // IRQ-0
        do RightFeeler() as interrupt 2; // IRQ-1
}

Robot:Init()
{
        // --------------------
        // Port-C initialization
        // Motor reversal
        //
        System:DDRC = System:PIN6_OUT
                    | System:PIN7_OUT;

        System:PORTC = 0;

        // -------------------
        // Heartbeat initialization
        //
        beat = PULSE_CNT;
```

```
        System:TCCR0 = System:T0_CK1024;      // Clock step once every 1024 cycles
        System:TIMSK = System:TOIE0;          // Enable Timer-0 overflow interrupt

        // ---------------------
        // Port-D initialization
        // Feelers, PWM, and Heart
        //
        System:DDRD = System:PIN2_IN                 // Left feeler
                    | System:PIN3_IN                 // Right feeler
                    | System:PIN4_OUT                // Heartbeat indicator
                    | System:PIN5_OUT;               // PWM Out

        System:PORTD = System:PIN2_UP | System:PIN3_UP;

        //--------------------------
        // Feeler interrupts
        //
        System:MCUCR = System:INT0_FALL
                     | System:INT1_FALL;

        System:GIMSK = System:INT0
                     | System:INT1;

        // --------------------------
        // Motor PWM initialization
        //
        left_vel = 0;
        right_vel = 0;

        System:TCCR1A = System:PWM_8
                      | System:PWMA_NORM
                      | System:PWMB_NORM;

        System:TCCR1B = System:T1_CK8;
}

Robot:Pulse()
{
        beat = beat - 1;
        if (beat == 0)
        {
                beat = PULSE_CNT;
                System:PORTD = System:PORTD ^ HEART_ON;

                if (left_vel < VEL_MAX)
                {
                        System:OCR1AL = left_vel;
                        left_vel = left_vel + VEL_ADD;
                }
                else
                {
                        System:PORTC = System:PORTC & ~(LEFT_REV | RIGHT_REV);
                }

                if (right_vel < VEL_MAX)
                {
                        System:OCR1BL = right_vel;
                        right_vel = right_vel + VEL_ADD;
                }
```

```
                else
                {
                        System:PORTC = System:PORTC & ~(LEFT_REV | RIGHT_REV);
                }
        }
}

Robot:LeftFeeler()
{
        left_vel = VEL_MAX/4;
        right_vel = VEL_MAX/2;
        System:PORTC = System:PORTC | LEFT_REV | RIGHT_REV;
}

Robot:RightFeeler()
{
        left_vel = VEL_MAX/2;
        right_vel = VEL_MAX/4;
        System:PORTC = System:PORTC | LEFT_REV | RIGHT_REV;
}

// ====================================================================
//      start
//
//      The main program does little more than initialize interrupts
//      and die — all of the work is done by the interrupt handlers.
//
start()
{
        outch 0x0d;
        out "(legs)";
        outch 0x0d;

        Robot:Init();
}
```

Motor.fuz

```
// ====================================================================
//
//              Motor.fuz
//
//      Motor control module, includes reverse and PWM control
//
//              Port B, Pin 0  Left Reverse
//              Port B, Pin 1  Right Reverse
//              Port D, Pin 5  Left PWM (OC1A)
//              OC1B                    Right PWM
//
//      Copyright (c) 1999 Edwin Wise and Simulated Reality Systems
//      http://www.simreal.com/Fuzbol
//
// ====================================================================


define Motor
{
        LeftVel as memory 0x4a;        // OCR1AL
        RightVel as memory 0x48;       // OCR1BL
```

```
        do Init();                          // Initialize PWM

        do LeftRev( rev as byte );
        do RightRev( rev as byte );
        do Stop();
}

Motor:Init()
{
        // ------------------------
        // Port-C initialization
        // Motor reversal
        //
        System:DDRB = System:DDRB
                    | System:PIN0_OUT
                    | System:PIN1_OUT;

        System:PORTB = System:PORTB &
                     ~(System:PIN0
                         | System:PIN1);


        // ------------------------
        // PWM initialization
        //
        System:DDRD = System:DDRD
                    | System:PIN5_OUT;                           // PWM Out

        System:TCCR1A = System:PWM_8
                      | System:PWMA_NORM
                      | System:PWMB_NORM;

        System:TCCR1B = System:T1_CK8;

        LeftVel= 0;
        RightVel = 0;
}


Motor:LeftRev( rev as byte )
{
        if (rev)
        {
                System:PORTB = System:PORTB | System:PIN0;
        }
        else
        {
                System:PORTB = System:PORTB & ~System:PIN0;
        }
}

Motor:RightRev( rev as byte )
{
        if (rev)
        {
                System:PORTB = System:PORTB | System:PIN1;
        }
        else
        {
```

```
                            System:PORTB = System:PORTB & ~System:PIN1;
            }
}


Motor:Stop()
{
        LeftVel = 0;
        RightVel = 0;

        LeftRev( 0 );
        RightRev( 0 );
}
```

ioPort.fuz

```
// ====================================================================
//
//              ioPort.fuz
//
//      I/O Port Control Module
//
//              0xD000                  Digital Input
//              0xE000                  Digital Output
//              0xF000                  Analog Input
//
//              Port D, Pin 3  Interrupt 1, analog conversion done
//
//      Copyright (c) 1999 Edwin Wise and Simulated Reality Systems
//      http://www.simreal.com/Fuzbol
//
// ====================================================================


define ioPort
{
        DigitalOut      as memory 0xD000;
        DigitalIn       as memory 0xE000;
        Analog as byte[8];                      // Storage of analog data

        a_to_d          as memory 0xF000;       // Do not use directly
        channel as byte;                        // Current conversion channel

        do Init();
        do Convert() as interrupt 2;
}


// ====================================================================
//
//      Sets the interrupt, and starts a conversion on channel 0
//
ioPort:Init()
{
        idx as integer;

        System:DDRD = System:DDRD
                                        & ~System:PIN3_IN;

        System:PORTD = System:PORTD
```

```
                                      |  System:PIN3_UP;

        System:MCUCR = System:MCUCR
                         |  System:INT1_FALL;

        System:GIMSK = System:GIMSK
                         |  System:INT1;


        channel = 0;
        a_to_d = (0x08 |  channel);
}


ioPort:Convert()
{
        //
        // Slow down memory and get results of this conversion
        //
        System:MCUCR = System:MCUCR |  System:SRW;

        Analog[channel] = a_to_d;

        //
        // Start the next conversion...
        //
        channel = (channel + 1) & 0x07;
        a_to_d = (0x08 |  channel);

        System:MCUCR = System:MCUCR & ~System:SRW;

}
```

Timid.fuz


```
// =====================================================================
//
//              Timid.fuz
//
//      Timid Behavior.
//
//              Port D, Pin4            Heartbeat Indicator
//
//      Copyright (c) 1999 Edwin Wise and Simulated Reality Systems
//      http://www.simreal.com/Fuzbol
//
// =====================================================================

use "system_8515.fuz";
use "ioPort.fuz";
use "Motor.fuz";


// =====================================================================

define Robot
{
        beat as integer;                         // Heartbeat counter
        bright as byte;                          // Pre-defined "brightness"
```

```
        const FULL_SPEED = 0x80;
        const FULL_STOP = 0x00;

        do Init();                              // Initialize Heartbeat timer
        do Pulse() as interrupt7;               // Interrupt handler on Timer-0
                const PULSE_CNT = 5;            // How many interrupts per pulse
                const ON = 0x10;                // OR to turn on
}

// ===================================================================

Robot:Init()
{
        beat = PULSE_CNT;
        bright = 0;

        System:DDRD = System:DDRD
                        | System:PIN4_OUT;
        System:PORTD = System:PORTD
                        & ~ON;

        System:TCNT0 = 0;                       // Init counter to 0
        System:TCCR0 = System:TCCR0
                        | System:T0_CK1024;     // Clock step once every 1024 cycles
        System:TIMSK = System:TIMSK
                        | System:TOIE0;         // Enable Timer-0 overflow interrupt
}

Robot:Pulse()
{
        left_eye as byte;
        right_eye as byte;

        beat = beat - 1;
        if (beat <= 0)
        {
                beat = PULSE_CNT;
                System:PORTD = System:PORTD ^ ON;

                left_eye = ioPort:Analog[0];
                right_eye = ioPort:Analog[1];

                if (bright == 0)
                {
                        bright = (left_eye max right_eye) + 4;
                }
                else
                {
                        if ( (left_eye > bright)
                                | (right_eye > bright) )
                        {
                                Motor:LeftVel = FULL_SPEED;
                                Motor:RightVel = FULL_SPEED;
                        }
                        else
                        {
                                Motor:LeftVel = FULL_STOP;
                                Motor:RightVel = FULL_STOP;
                        }
```

```
                }
            }
}


// ===================================================================
//      start
//
start()
{
        outch 0x0d;
        out "Timid";
        outch 0x0d;

        Robot:Init();
        ioPort:Init();
        Motor:Init();
        Motor:Stop();
}
```

Fear.fuz

```
// ===================================================================
//
//              Fear.fuz
//
//      Fearful Behavior — photo-phobe
//
//              Port D, Pin4            Heartbeat Indicator
//
//      Copyright (c) 1999 Edwin Wise and Simulated Reality Systems
//      http://www.simreal.com/Fuzbol
//
// ===================================================================
use "system_8515.fuz";
use "ioPort.fuz";
use "Motor.fuz";


// ===================================================================

range BLACK    ( 0x00, 0x00, 0x10 );
range DARK     ( 0x20, 0x00, 0x18 );
range DIM      ( 0x40, 0x00, 0x20 );
range BRIGHT( 0x90, 0x00, 0x50 );
range LIGHT    ( 0xF0, 0x10, 0x50 );

rule MoveLight
{
        BLACK    as 0x00;
        DARK     as 0xA0;
        DIM              as 0xC0;
        BRIGHT   as 0xF0;
        LIGHT    as 0xFF;
}

// ===================================================================
```

```
define Robot
{
        beat as integer;                              // Heartbeat counter
        dark as byte;

        const FULL_SPEED = 0x80;
        const FULL_STOP = 0x00;

        do Init();                            // Initialize Heartbeat timer
        do Pulse() as interrupt7;             // Interrupt handler on Timer-0
                const PULSE_CNT = 5;          // How many interrupts per pulse
                const ON = 0x10;              // OR to turn on
}

// ====================================================================

Robot:Init()
{
        beat = PULSE_CNT;
        dark = 0;

        System:DDRD = System:DDRD
                        | System:PIN4_OUT;
        System:PORTD = System:PORTD
                                    & ~ON;

        System:TCNT0 = 0;                     // Init counter to 0
        System:TCCR0 = System:TCCR0
                        | System:T0_CK1024;   // Clock step once every 1024 cycles
        System:TIMSK = System:TIMSK
                        | System:TOIE0;       // Enable Timer-0 overflow interrupt
}

Robot:Pulse()
{
        left_eye as byte;
        right_eye as byte;

        beat = beat - 1;
        if (beat <= 0)
        {
                beat = PULSE_CNT;
                System:PORTD = System:PORTD ^ ON;

                left_eye = ioPort:Analog[0];
                right_eye = ioPort:Analog[1];

                if (dark == 0)
                {
                        //
                        // At first time through, decide what "dark" is
                        //
                        dark = (left_eye min right_eye);
                        if (dark > 4)
                        {
                                dark = dark - 4;
                        }
                }
                else
                {
```

```
            //
            // Calibrate the current eye values relative to
            //   relative "darkness"
            //
            if (left_eye < dark)
            {
                left_eye = 0;
            }
            else
            {
                left_eye = left_eye - dark;
            }
            if (right_eye < dark)
            {
                right_eye = 0;
            }
            else
            {
                right_eye = right_eye - dark;
            }

            //
            // Run away from the light
            // (actuate the motor near the light proportional
            // to the light level)
            //
            Motor:LeftVel = left_eye is MoveLight;
            Motor:RightVel = right_eye is MoveLight;
        }
    }
}


// ====================================================================
//      start
//
start()
{
        outch 0x0d;
        out "Fear";
        outch 0x0d;

        Robot:Init();
        ioPort:Init();
        Motor:Init();
        Motor:Stop();
}
```

Aggression.fuz

```
// ====================================================================
//
//
//            Aggression.fuz
//
//      Aggressive Behavior — photo-vore, to an extreme
//
//            Port D, Pin4           Heartbeat Indicator
//
```

```
//      Copyright (c) 1999 Edwin Wise and Simulated Reality Systems
//      http://www.simreal.com/Fuzbol
//
// ==================================================================

use "system_8515.fuz";
use "ioPort.fuz";
use "Motor.fuz";


// ==================================================================

range BLACK    ( 0x00, 0x00, 0x20 );
range DARK     ( 0x40, 0x00, 0x30 );
range DIM      ( 0x80, 0x00, 0x30 );
range BRIGHT( 0xC0, 0x00, 0x30 );
range LIGHT    ( 0xFF, 0x00, 0x20 );

rule MoveLight
{
        BLACK    as 0x40;
        DARK     as 0x80;
        DIM      as 0xC0;
        BRIGHT   as 0xE0;
        LIGHT    as 0xFF;
}


// ==================================================================

define Robot
{
        beat as integer;                    // Heartbeat counter
        dark as byte;

        const FULL_SPEED = 0x80;
        const FULL_STOP = 0x00;

        do Init();                          // Initialize Heartbeat timer
        do Pulse() as interrupt 7;          // Interrupt handler on Timer-0
                const PULSE_CNT = 5;        // How many interrupts per pulse
                const ON = 0x10;            // OR to turn on
}


// ==================================================================

Robot:Init()
{
        beat = PULSE_CNT;
        dark = 0;

        System:DDRD = System:DDRD
                        | System:PIN4_OUT;
        System:PORTD = System:PORTD
                        & ~ON;

        System:TCNT0 = 0;                   // Init counter to 0
        System:TCCR0 = System:TCCR0
                        | System:T0_CK1024; // Clock step once every 1024 cycles
        System:TIMSK = System:TIMSK
                        | System:TOIE0;     // Enable Timer-0 overflow interrupt
}
```

```
Robot:Pulse()
{
        left_eye as byte;
        right_eye as byte;

        beat = beat - 1;
        if (beat <= 0)
        {
                beat = PULSE_CNT;
                System:PORTD = System:PORTD ^ ON;

                left_eye = ioPort:Analog[0];
                right_eye = ioPort:Analog[1];

                if (dark == 0)
                {
                        //
                        // At first time through, decide what "dark" is
                        //
                        dark = (left_eye min right_eye);
                        if (dark > 4)
                        {
                                dark = dark - 4;
                        }
                }
                else
                {
                        //
                        // Calibrate the current eye values relative to
                        //      relative "darkness"
                        //
                        if (left_eye < dark)
                        {
                                left_eye = 0;
                        }
                        else
                        {
                                left_eye = left_eye - dark;
                        }
                        if (right_eye < dark)
                        {
                                right_eye = 0;
                        }
                        else
                        {
                                right_eye = right_eye - dark;
                        }

                        //
                        // Run towards the light
                        //
                        Motor:RightVel = left_eye is MoveLight;
                        Motor:LeftVel = right_eye is MoveLight;
                }
        }
}

// ====================================================================
```

```
//      start
//
//
start()
{
        outch 0x0d;
        out "Aggression";
        outch 0x0d;

        Robot:Init();
        ioPort:Init();
        Motor:Init();
        Motor:Stop();
}
```

Love.fuz

```
// ====================================================================
//
//              Love.fuz
//
//              Port D, Pin4            Heartbeat Indicator
//
//      Copyright (c) 1999 Edwin Wise and Simulated Reality Systems
//      http://www.simreal.com/Fuzbol
//
// ====================================================================

use "system_8515.fuz";
use "ioPort.fuz";
use "Motor.fuz";

// ====================================================================

range BLACK    ( 0x00, 0x00, 0x10 );
range DARK     ( 0x20, 0x00, 0x18 );
range DIM      ( 0x40, 0x00, 0x20 );
range BRIGHT( 0x90, 0x00, 0x50 );
range LIGHT    ( 0xF0, 0x10, 0x50 );

rule MoveLight
{
        BLACK   as 0x00;
        DARK    as 0xA0;
        DIM             as 0xC0;
        BRIGHT  as 0xF0;
        LIGHT   as 0xFF;
}

// ====================================================================

define Robot
{
        beat as integer;                // Heartbeat counter
        dark as byte;

        const FULL_SPEED = 0x80;
        const FULL_STOP = 0x00;
```

```
        do Init();                    // Initialize Heartbeat timer
        do Pulse() as interrupt 7;    // Interrupt handler on Timer-0
              const PULSE_CNT = 5;            // How many interrupts per pulse
              const ON = 0x10;               // OR to turn on
}


// =================================================================

Robot:Init()
{
        beat = PULSE_CNT;
        dark = 255;

        System:DDRD = System:DDRD
                        | System:PIN4_OUT;
        System:PORTD = System:PORTD
                        & ~ON;

        System:TCNT0 = 0;                     // Init counter to 0
        System:TCCR0 = System:TCCR0
                        | System:T0_CK1024;   // Clock step once every 1024 cycles
        System:TIMSK = System:TIMSK
                        | System:TOIE0;       // Enable Timer-0 overflow interrupt
}

Robot:Pulse()
{
        left_eye as byte;
        right_eye as byte;

        beat = beat - 1;
        if (beat <= 0)
        {
                beat = PULSE_CNT;
                System:PORTD = System:PORTD ^ ON;

                left_eye = ioPort:Analog[0];
                right_eye = ioPort:Analog[1];

                //
                // Continuously update dark
                //
                dark = dark min (left_eye min right_eye);

                //
                // Calibrate the current eye values relative to
                //      relative "darkness"
                //
                left_eye = left_eye - dark;
                right_eye = right_eye - dark;

                //
                // Wander until the light is found
                //
                Motor:LeftVel = 255 - (left_eye is MoveLight);
                Motor:RightVel = 255 - (right_eye is MoveLight);
        }
}
```

```
// ====================================================================
//      start
//
//
start()
{
        outch 0x0d;
        out "Love";
        outch 0x0d;

        Robot:Init();
        ioPort:Init();
        Motor:Init();
        Motor:Stop();
}
```

Dogged.fuz

```
// ====================================================================
//
//              Dogged.fuz
//
//              Port D, Pin 2           Interrupt 0
//              Port D, Pin 4           Heartbeat Indicator
//
//      Copyright (c) 1999 Edwin Wise and Simulated Reality Systems
//      http://www.simreal.com/Fuzbol
//
// ====================================================================

use "system_8515.fuz";
use "ioPort.fuz";
use "Motor.fuz";

// ====================================================================

range BLACK    ( 0x00, 0x00, 0x10 );
range DARK     ( 0x20, 0x00, 0x18 );
range DIM      ( 0x40, 0x00, 0x20 );
range BRIGHT( 0x90, 0x00, 0x50 );
range LIGHT    ( 0xF0, 0x10, 0x50 );

rule MoveLight
{
        BLACK   as 0x00;
        DARK    as 0x90;
        DIM             as 0xB0;
        BRIGHT  as 0xE0;
        LIGHT   as 0xFF;
}

// ====================================================================

define Robot
{
        beat as integer;                        // Heartbeat counter
        dark as byte;                           // Dark level
```

```
            // Robot's status...
            left_go as integer;                   // Target velocity
            right_go as integer;

            left_rev as byte;                     // Reverse control
            right_rev as byte;
            count_rev as byte;

            const FULL_SPEED = 0x80;
            const FULL_STOP = 0x00;

            do Init();                            // Initialize Heartbeat timer

            do Pulse() as interrupt 7;            // Interrupt handler on Timer-0
                    const PULSE_CNT = 5;          // How many interrupts per pulse
                    const ON = 0x10;              // OR to turn on

            do Bump() as interrupt 1;
            do Seek( left_eye as byte, right_eye as byte );
            do Reverse( feelers as byte );
            do Update();
}

// =================================================================

Robot:Init()
{
            beat = PULSE_CNT;
            dark = 255;
            count_rev = 0;

            //
            // Heartbeat light
            //
            System:DDRD = System:DDRD
                            | System:PIN4_OUT;
            System:PORTD = System:PORTD
                            & ~ON;
            //
            // Heartbeat timer and interrupt
            //
            System:TCNT0 = 0;                     // Init counter to 0
            System:TCCR0 = System:TCCR0
                            | System:T0_CK1024;   // Clock step once every 1024 cycles
            System:TIMSK = System:TIMSK
                            | System:TOIE0;       // Enable Timer-0 overflow interrupt

            //
            // Feeler interrupt
            //
            System:MCUCR = System:MCUCR
                            | System:INT0_FALL;

            System:GIMSK = System:GIMSK
                            | System:INT0 ;

}

Robot:Pulse()
```

```
{
        left_eye as byte;
        right_eye as byte;

        beat = beat - 1;
        if (beat <= 0)
        {
                beat = PULSE_CNT;
                System:PORTD = System:PORTD ^ ON;

                left_eye = ioPort:Analog[0];
                right_eye = ioPort:Analog[1];

                //
                // Continuously update dark
                //
                dark = dark min (left_eye min right_eye);

                //
                // Calibrate the current eye values relative to
                //       relative "darkness"
                //
                left_eye = left_eye - dark;
                right_eye = right_eye - dark;

                if (count_rev == 0)
                {
                        //
                        // Only seek light when not escaping obstacles
                        //
                        Seek( left_eye, right_eye );
                }
                Reverse( ioPort:DigitalIn & 0x03 );
                Update();
        }
}

Robot:Bump()
{
        //
        // On bump, come to a dead stop
        //
        if (count_rev == 0)
        {
                Motor:LeftVel = 0;
                Motor:RightVel = 0;

                left_go = 0;
                right_go = 0;
        }
}

Robot:Seek(
        left_eye as byte,
        right_eye as byte )
{
        //
        // Wander until the light is found
        //
        left_go = 255 - (left_eye is MoveLight);
```

```
        right_go = 255 - (right_eye is MoveLight);

        left_rev = 0;
        right_rev = 0;
}

Robot:Reverse(
        feelers as byte )
{
        //
        // When in reverse mode, stick to it until done
        //
        if (count_rev > 0)
        {
                count_rev = count_rev - 1;
        }
        else
        {
                if (feelers == 0x03)
                {
                        // Both hit, straight reverse
                        //(rarely occurs; one will almost always make
                        //contact before the other)
                        //
                        left_rev = 1;
                        right_rev = 1;

                        left_go = 0x70;
                        right_go = 0x80;

                        count_rev = 20;
                }
                elseif (feelers == 0x02)
                {
                        // Right feeler hit
                        left_rev = 1;
                        right_rev = 1;

                        left_go = 0x40;
                        right_go = 0x80;

                        count_rev = 15;
                }
                elseif (feelers == 0x01)
                {
                        // Left feeler hit
                        // Right feeler hit
                        left_rev = 1;
                        right_rev = 1;

                        left_go = 0x80;
                        right_go = 0x40;

                        count_rev = 15;
                }
        }
}

Robot:Update()
{
```

```
        Motor:LeftVel = (Motor:LeftVel + left_go) >> 1;
        Motor:RightVel = (Motor:RightVel + right_go) >> 1;

        Motor:LeftRev( left_rev );
        Motor:RightRev( right_rev );
}




// ====================================================================
//      start
//
start()
{
        outch 0x0c;
        outch 0x0d;
        outch 0x0d;
        out "Dogged";
        outch 0x0d;

        Robot:Init();
        ioPort:Init();
        Motor:Init();
        Motor:Stop();
}
```

Insecure.fuz

```
// ====================================================================
//
//              Insecure.fuz
//
//              Port D, Pin 2           Interrupt 0
//              Port D, Pin 4           Heartbeat Indicator
//
//      Copyright (c) 1999 Edwin Wise and Simulated Reality Systems
//      http://www.simreal.com/Fuzbol
//
// ====================================================================

use "system_8515.fuz";
use "ioPort.fuz";
use "Motor.fuz";

// ====================================================================

range BLACK    ( 0x00, 0x00, 0x10 );
range DARK     ( 0x20, 0x00, 0x18 );
range DIM      ( 0x40, 0x00, 0x20 );
range BRIGHT( 0x90, 0x00, 0x50 );
range LIGHT    ( 0xF0, 0x10, 0x50 );

range HAPPY ( 0x00, 0x20, 0xD0 );
range LOST     ( 0xA0, 0x00, 0x60 );
range PANIC    ( 0xF0, 0x10, 0x70 );

rule MoveLight
{
```

```
            BLACK    as 0x20;
            DARK     as 0x90;
            DIM             as 0xC0;
            BRIGHT   as 0xE0;
            LIGHT    as 0xFF;
}

// ===================================================================

define Robot
{
            beat as integer;                    // Heartbeat counter
            dark as byte;                       // Dark level
            security as byte;                   // Security counter

            // Robot's status...
            left_go as fuzzy;                   // Target velocity
            right_go as fuzzy;

            left_rev as byte;                   // Reverse control
            right_rev as byte;
            count_rev as byte;

            do Init();                          // Initialize Heartbeat timer

            do Pulse() as interrupt 7;          // Interrupt handler on Timer-0
                    const PULSE_CNT = 10;       // How many interrupts per pulse
                    const ON = 0x10;            // OR to turn on

            do Bump() as interrupt 1;
            do SeekLight( left_eye as byte, right_eye as byte );
            do SeekWall();
            do Panic();
            do Reverse( feelers as byte );
            do Update();
}

// ===================================================================

Robot:Init()
{
            beat = PULSE_CNT;
            dark = 255;
            security = 0;
            count_rev = 0;

            //
            // Heartbeat light
            //
            System:DDRD = System:DDRD
                        | System:PIN4_OUT;
            System:PORTD = System:PORTD
                        & ~ON;
            //
            // Heartbeat timer and interrupt
            //
            System:TCNT0 = 0;                   // Init counter to 0
            System:TCCR0 = System:TCCR0
                        | System:T0_CK1024;     // Clock step once every 1024 cycles
            System:TIMSK = System:TIMSK
```

```
                              | System:TOIE0;         // Enable Timer-0 overflow interrupt

        //
        // Feeler interrupt
        //
        System:MCUCR = System:MCUCR
                        | System:INT0_FALL;

        System:GIMSK = System:GIMSK
                        | System:INT0 ;

}

Robot:Pulse()
{
        left_eye as byte;
        right_eye as byte;

        beat = beat - 1;
        if (beat <= 0)
        {
                beat = PULSE_CNT;
                System:PORTD = System:PORTD ^ ON;

                left_eye = ioPort:Analog[0];
                right_eye = ioPort:Analog[1];

                //
                // Continuously update dark
                //
                dark = dark min (left_eye min right_eye);

                //
                // Calibrate the current eye values relative to
                //      relative "darkness"
                //
                left_eye = left_eye - dark;
                right_eye = right_eye - dark;

                //
                // Manage the security blanket
                //
                if (security < 255)
                {
                        security = security + 1;
                }

                //
                // Behaviors!
                //
                Reverse( ioPort:DigitalIn & 0x03 );
                if (count_rev == 0)
                {
                        clear( left_go );
                        clear( right_go );

                        left_rev = 0;
                        right_rev = 0;

                        SeekLight( left_eye, right_eye );
```

```
                        SeekWall();
                        Panic();
                }
                Update();
        }
}

Robot:Bump()
{
        //
        // On bump, come to a dead stop
        //
        if (count_rev == 0)
        {
                Motor:LeftVel = 0;
                Motor:RightVel = 0;

                clear( left_go );
                clear( right_go );

                //
                // Ahhh... a wall....
                //
                security = 0;
        }
}

Robot:SeekLight(
        left_eye as byte,
        right_eye as byte )
{
        weight( security is HAPPY );
        //
        // Wander until the light is found
        //
        left_go = right_eye is MoveLight;
        right_go = left_eye is MoveLight;
}

Robot:SeekWall()
{
        weight( security is LOST );

        //
        // When feeling lost, curve to the right...
        //
        left_go = 0xF0;
        right_go = 0x40;
}

Robot:Panic()
{
        weight( security is PANIC );

        //
        // When feeling REALLY lost, zip straight ahead!
        //
        left_go = 0xFF;
        right_go = 0xFF;
}
```

```
Robot:Reverse(
        feelers as byte )
{

        weight( 255 );

        //
        // When in reverse mode, stick to it until done
        //
        if (count_rev > 0)
        {
            count_rev = count_rev - 1;
        }
        else
        {
                if (feelers == 0x03)
                {
                        // Both hit, straight reverse
                        // (rarely occurs; one will almost always make
                        //    contact before the other)
                        //
                        left_rev = 1;
                        right_rev = 1;

                        left_go = 0x70;
                        right_go = 0x80;

                        count_rev = 5;
                }
                elseif (feelers == 0x02)
                {
                        // Right feeler hit
                        left_rev = 1;
                        right_rev = 1;

                        left_go = 0x80;
                        right_go = 0x40;

                        count_rev = 5;
                }
                elseif (feelers == 0x01)
                {
                        // Left feeler hit
                        // Right feeler hit
                        left_rev = 1;
                        right_rev = 1;

                        left_go = 0x40;
                        right_go = 0x80;

                        count_rev = 5;
                }
        }
}

Robot:Update()
{
        left_vel as byte;
        right_vel as byte;
```

```
        left_vel = left_go;
        right_vel = right_go;

        Motor:LeftVel = left_vel;
        Motor:RightVel = right_vel;

        Motor:LeftRev( left_rev );
        Motor:RightRev( right_rev );
}

// ==================================================================
//      start
//
start()
{
        outch 0x0d;
        out "Insecure";
        outch 0x0d;

        Robot:Init();
        ioPort:Init();
        Motor:Init();
        Motor:Stop();
}
```

Moth.fuz

```
// ==================================================================
//
//              Moth.fuz
//
//              Port D, Pin 2           Interrupt 0
//              Port D, Pin 4           Heartbeat Indicator
//
//      Copyright (c) 1999 Edwin Wise and Simulated Reality Systems
//      http://www.simreal.com/Fuzbol
//
// ==================================================================

use "system_8515.fuz";
use "ioPort.fuz";
use "Motor.fuz";

// ==================================================================

range BLACK    ( 0x00, 0x00, 0x40 );
range DARK     ( 0x40, 0x00, 0x40 );
range DIM      ( 0x80, 0x00, 0x40 );
range BRIGHT( 0xC0, 0x00, 0x40 );
range LIGHT    ( 0xFF, 0x00, 0x40 );

rule MoveLight
{
        BLACK   as 0x40;
        DARK    as 0x98;
        DIM             as 0xC0;
        BRIGHT  as 0xFF;
        LIGHT   as 0x00;
```

```
}

// ====================================================================

define Robot
{
        beat as integer;                    // Heartbeat counter
        dark as byte;                       // Dark level
        light as byte;

        // Robot's status...
        left_go as byte;                    // Target velocity
        right_go as byte;
        left_vel as integer;
        right_vel as integer;

        left_rev as byte;                   // Reverse control
        right_rev as byte;
        count_rev as byte;

        do Init();                          // Initialize Heartbeat timer

        do Pulse() as interrupt 7;          // Interrupt handler on Timer-0
                const PULSE_CNT = 10;       // How many interrupts per pulse
                const ON = 0x10;            // OR to turn on

        do Bump() as interrupt 1;
        do SeekLight( left_eye as byte, right_eye as byte );
        do Reverse( feelers as byte );
        do Update();
}

// ====================================================================

Robot:Init()
{
        beat = PULSE_CNT;
        dark = 0xFF;
        light = 0x00;
        count_rev = 0;

        left_vel = 0;
        right_vel = 0;

        //
        // Heartbeat light
        //
        System:DDRD = System:DDRD
                        | System:PIN4_OUT;
        System:PORTD = System:PORTD
                        & ~ON;
        //
        // Heartbeat timer and interrupt
        //
        System:TCNT0 = 0;                   // Init counter to 0
        System:TCCR0 = System:TCCR0
                        | System:T0_CK1024; // Clock step once every 1024 cycles
        System:TIMSK = System:TIMSK
                        | System:TOIE0;     // Enable Timer-0 overflow interrupt
```

```
        //
        // Feeler interrupt
        //
        System:MCUCR = System:MCUCR
                        | System:INT0_FALL;

        System:GIMSK = System:GIMSK
                        | System:INT0 ;


}

Robot:Pulse()
{
        left_eye as byte;
        right_eye as byte;
        scale as real;

        beat = beat - 1;
        if (beat <= 0)
        {

        beat = PULSE_CNT;
        System:PORTD = System:PORTD ^ ON;

        left_eye = ioPort:Analog[0];
        right_eye = ioPort:Analog[1];

        //
        // Continuously update dark
        //
        dark = dark min (left_eye min right_eye);
        light = light max (left_eye max right_eye);
        scale = 255.0 / (light-dark);
        if (scale < 1.0)
        {
                scale = 1.0;
        }

        //
        // Calibrate the current eye values relative to
        //      relative "darkness"
        //
        left_eye = (left_eye - dark) * scale;
        right_eye = (right_eye - dark) * scale;

        //
        // Behaviors!
        //
        Reverse( ioPort:DigitalIn & 0x03 );
        if (count_rev == 0)
        {
                left_rev = 0;
                right_rev = 0;
                SeekLight( left_eye, right_eye );
        }
        Update();
        }
}

Robot:Bump()
```

```
{
        //
        // On bump, come to a dead stop
        //
        if (count_rev == 0)
        {
                left_vel = 0;
                right_vel = 0;

                Motor:LeftVel = 0;
                Motor:RightVel = 0;
        }
}

Robot:SeekLight(
        left_eye as byte,
        right_eye as byte )
{
        left_go = right_eye is MoveLight;
        right_go = left_eye is MoveLight;
}

Robot:Reverse(
        feelers as byte )
{
        weight( 255 );

        //
        // When in reverse mode, stick to it until done
        //
        if (count_rev > 0)
        {
                count_rev = count_rev - 1;
        }
        else
        {
                if (feelers == 0x03)
                {
                        // Both hit, straight reverse
                        // (rarely occurs; one will almost always make
                        //      contact before the other)
                        //
                        left_rev = 1;
                        right_rev = 1;

                        left_go = 0x70;
                        right_go = 0x80;

                        count_rev = 5;
                }
                elseif (feelers == 0x02)
                {
                        // Right feeler hit
                        left_rev = 1;
                        right_rev = 1;

                        left_go = 0x80;
                        right_go = 0x40;

                        count_rev = 5;
```

```
                }
                elseif (feelers == 0x01)
                {
                        // Left feeler hit
                        // Right feeler hit
                        left_rev = 1;
                        right_rev = 1;

                        left_go = 0x40;
                        right_go = 0x80;

                        count_rev = 5;
                }
        }
}

Robot:Update()
{
        left_vel = (left_vel + left_go) >> 1;
        right_vel = (right_vel + right_go) >> 1;

        Motor:LeftVel = left_vel;
        Motor:RightVel = right_vel;

        Motor:LeftRev( left_rev );
        Motor:RightRev( right_rev );
}

// ====================================================================
//      start
//
start()
{
        outch 0x0d;
        out "Moth";
        outch 0x0d;

        Robot:Init();
        ioPort:Init();
        Motor:Init();
        Motor:Stop();
}
```

Servo.fuz

```
// ====================================================================
//
//              servo.fuz
//
//      Copyright (c) 1999 Edwin Wise and Simulated Reality Systems
//      http://www.simreal.com/Fuzbol
//
// ====================================================================

use "system_8515.fuz";
use "ioPort.fuz";


// ====================================================================
```

```
define Servo
{
        time as byte[16];
        beat as byte;
        port as byte;
        signal as byte;

        do Init();                              // Initialize Heartbeat timer

        do Pulse() as interrupt       7;        // Interrupt handler on Timer-0
                const ON = 0x10;
}

// ===================================================================

Servo:Init()
{
        idx as byte;
        tmp as byte;
        pos as byte;

        idx = 0;
        pos = 0;
        while (idx < 16)
        {
                time[idx]     = 128 + pos;
                time[idx+1] = 255 - pos;
                idx = idx + 2;

                pos = pos + 16;
        }

        //
        // Heartbeat light
        //
        System:DDRD = System:DDRD
                        | System:PIN4_OUT;
        System:PORTD = System:PORTD
                        & ~ON;
        //
        // Heartbeat timer and interrupt
        //
        System:TCNT0 = 0;
        System:TCCR0 = System:TCCR0
                        | System:T0_CK64;
        System:TIMSK = System:TIMSK
                        | System:TOIE0;

        beat = 0;
        port = 1;
        signal = 0xff;
}

Servo:Pulse()
{
        System:TCNT0 = 255 - time[beat];
        if (beat & 0x01)
        {
                signal = signal & ~port;
                port = port << 1;
```

```
            }
            else
            {
                    signal  = signal | port;
            }
            ioPort:DigitalOut = signal;
            beat = beat + 1;

            if (beat >= 16)
            {
                    idx as byte;
                    val as byte;

                    beat = 0;
                    System:PORTD = System:PORTD ^ ON;

                    port = 0;
                    idx = 0;
                    while (idx < 16)
                    {
                            val = ioPort:Analog[port] >> 1;

                            time[idx] = 128 + val;
                            time[idx+1] = 64 + (128 - val);

                            port = port + 1;
                            idx = idx + 2;
                    }
                    port = 1;
            }
}




// ====================================================================
//      start
//
start()
{
        outch 0x0c;
        outch 0x0d;
        outch 0x0d;
        out "Servo";
        outch 0x0d;

        ioPort:Init();
        Servo:Init();
}
```

Appendix C: Conversions and Code References

Units

Prefix	Power of 10	Decimal	Example (in Seconds)
pico- (p)	10e-12	.000,000,000,000,1	1,000,000,000,000 pS = 1S
nano- (n)	10e-9	.000,000,000,1	1,000,000,000 nS = 1S
micro- (u, μ)	10e-6	.000,000,1	1,000,000 uS = 1S
milli- (m)	10e-3	.000,1	1,000 mS = 1S
kilo- (K)	10e+3	1,000	1KΩ = 1,000Ω
meg- (M)	10e+6	1,000,000	1MΩ = 1,000,000Ω

A	Ampere
F	Farad
I	Current (amperes), I = V/R
P	Power (watts), P = VI = RI^2)
R	Resistance (ohms), R = V/I
V or E	Voltage, V = IR
W	Watt
Ω	Ohm

Resistor Codes

The first three bands on the resistor give the value in Ohms (Ω), as shown below. The color of the third band indicates how many zeros to add after the first two digits. An optional tolerance color, which is metallic and usually hard to see, gives the resistor's accuracy as a percent of its total value. Resistor color bands are shown in *Figure B-1*.

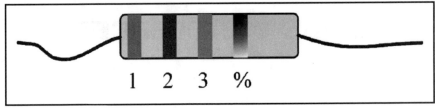

Figure B-1. Resistor color bands.

Color	First Band	Second Band	Third Band	(Tolerance)
Black	0	0	* 1	
Brown	1	1	* 10	
Red	2	2	* 100	
Orange	3	3	* 1,000	
Yellow	4	4	* 10,000	
Green	5	5	* 100,000	
Blue	6	6	* 1,000,000	
Violet	7	7	* 10,000,000	
Gray	8	8	* 100,000,000	
White	9	9		
Gold				+- 5%
Silver				+- 10%
(no metal)				+- 20%

Capacitor Codes

Capacitors will often have their capacitance value printed on their package. It is also common, for smaller capacitors, for their value to be encoded. This code consists of two or three digits:

XX or XXY

The first two digits XX are the numerical value of the capacitor in picoFarads, and Y (if it is given) is the power-of-ten multiplier. For example:

18	18 x 10^0 picofarads = 18pf
103	10 x 10^3 picofarads = 10,000pf = .01uf
475	47 x 10^5 picofarads = 4,700,000pf = 4.7uf

Suppliers

Simulated Reality Systems (SRS) Atmel MCU and Fuzbol systems	Book errata, software updates, project supplies. http://www.simreal.com Contact the author at: ewise@simreal.com
Alberta Printed Circuits, Ltd Printed Circuit board Prototyping	Unit 3, 1112-40th Ave NE Calgary, AB,Canada T2E 5T8 1 (403) 250-3406 http://www.apcircuits.com/
Arrick Robotics Robotics Specialty Store	P.O. Box 1574 Hurst, Texas, 76053 USA 1 (817) 571-4528 http://www.robotics.com/index.html
Clippard Instrument Laboratory, Inc. Pneumatic Components	http://www.clippard.com/
Digi-Key Electronic Components	701 Brooks Avenue South Thief River Falls, MN 56701-2757 USA 1 (800) 344-4539 http://www.digikey.com/
Humphrey Pneumatic Components	P.O. Box 2008 Kalamazoo MI 49003 USA 1 (800) 477-8707 http://www.humphreypc.com/
InnoMedia, Inc. Wireless Modems	90 Rio Robles Suite 100 San Jose, CA 95134 USA 1 (408) 432-5400 http://www.innomedia.com/

Marshall Electronics Electronic Components	9320 Telstar Avenue El Monte, CA 91731-2895 USA 1 (800) 833-9910 http://www.marshall.com/
Mouser Electronics Electronic Components	1 (800) 346-6875 http://www.mouser.com
Newark Electronics Electronic Components	1 (800) 4-Newark (463-9275) http://www.newark.com/
Parken-Hannifin Pneumatic Components	1 (800) C-Parker (272-7537) http://www.parker.com/
Pisco USA, Inc Pneumatic Fittings	2228 Landmeier Rd. Elk Grove Village, IL 60007 USA 1 (847) 427-1314 http://www.pisco.com/
Radio Shack Electronic Components	Found all over the world… 1 (800) THE SHACK (843-7812) http://www.radioshack.com/
Robot Store (Mondo-tronics) Robotics Specialty Store	1 (800) 374-5764 http://www.robotstore.com
Scott Edwards Electronics	1939 S. Frontage Rd., Suite F
LCD Displays	Sierra Vista AZ 85635 USA 1 (520) 459-4802 http://www.seetron.com
Small Parts, Inc. Mechanical Parts	1 (800) 220-4242 http://www.smallparts.com
Tech America Electronic Components	1 (800) 877-0072 http://www.techamerica.com/
W.W. Grainger, Inc. Mechanical and Industrial Components	455 Knightsbridge Parkway Lincolnshire, IL 60069-3639 USA http://www.grainger.com/

| Wirz Electronics
Hobbyist and Robotics supplies | P.O. Box 457
Littleton, MA 01460-0457 USA
1 (888) 289-9479
http://www.wirz.com/ |

Bibliography

Arkin, Ronald C. *Behavior Based Robotics.* MIT Press, 1998

Borenstein, J., et.al. *Navigating Mobile Robots: Systems and Techniques.* A.K. Peters, Ltd, 1996

Braitenberg, Valentino. *Vehicles: Experiments in Synthetic Psychology.* MIT Press, 1986

Brooks, Richard R., and Iyengar, S.S. *Multi-Sensor Fusion.* Prentice-Hall, 1998

Carr, Joseph J. *Electronic Circuit Guidebook Volume 1: Sensors.* PROMPT Publications, 1997

Carr, Joseph J. *Electronic Circuit Guidebook Volume 3: Op Amps.* PROMPT Publications, 1997

Caudill, Maureen, and Butler, Charles. *Naturally Intelligent Systems.* MIT Press, 1990

Everett, H.R. *Sensors for Mobile Robots: Theory and Application.* A.K. Peters, Ltd, 1998

Horn, Delton T. *Basic Electronics Theory.* 4th Edition. TAB Books, 1994

Horowitz, Paul, and Hill, Winfield. *The Art of Electronics.* 2nd Edition. Cambridge University Press, 1989

Iovine, John. *Robots, Androids, and Animatrons.* McGraw-Hill, 1997

Jones, Joseph and Flynn, Anita. *Mobile Robots – Inspiration to Implementation.* 1st Edition. A.K. Peters, Ltd, 1993

Kosko, Bart. *Fuzzy Engineering.* Prentice-Hall, 1997

Kosko, Bart. *Fuzzy Thinking: The New Science of Fuzzy Logic.* Hyperion, 1994

Lancaster, Don. *TTL Cookbook.* SAMS, 1974

Lenk, John D. *Simplified Design of Micropower and Battery Circuits.* Butterworth-Heinemann, 1996

Ross, J.N. *The Essence of Power Electronics.* Prentice-Hall, 1997

Sutton, Richard S., and Barto, Andrew G. *Reinforcement Learning.* MIT Press, 1998

Index

Semiconductor Cross Reference Book/CD-ROM

Sams Technical Publishing

Sams Technical Publishing has added thousands of new semiconductors to the fifth edition of the book and the second edition of its CD-ROM version. It is the most comprehensive cross-reference available for engineers, technicians, and all those who work with semiconductors. Parts from ECG, NTE, TCE, and Radio Shack are included along with coverage of all major semiconductor types: bipolar transistors, FETs, diodes, rectifiers, ICs, SCRs, LEDs, modules, and thermal devices. Key Features include:

- Complete guide to semiconductor replacement.
- An excellent tool for technicians of any experience level.
- More than 628,000 semiconductors listed.
- Up-to-date list of original equipment manufacturers.

Professional Reference
876 pages • paperback • 10 7/8 x 8 3/8"
ISBN 0-7906-1139-2 • Sams 61139
$39.95

Professional Reference
CD-ROM
ISBN 0-7906-1231-3 • Sams 61231
$39.95

Servicing TV/VCRCombo Units

Homer Davidson

Part of Sams Technical Publishing's Servicing Series, this comprehensive resource covers the repair issues surrounding the popular TV/VCR combo units. These electronic devices have become smaller, more affordable and functional. Thus, they are now being used in new ways, such as in autos, campers, kitchens, and other nontraditional locations.

TV/VCR combo units are serviced more often than other types of TVs, so a thorough understanding of their inner workings is essential to technicians and repair persons. *Servicing TV/VCR Combo Units* provides "tips and tricks" related to the equipment vs. specific models. It also goes into detail regarding popular models, with advice on how to apply repair techniques to other manufacturers' sets.

Author Homer Davidson has owned and operated his own TV repair shop in Fort Dodge, IA for 38 years. In *Servicing TV/VCR Combo Units*, Davidson puts his vast electronics knowledge to the task of tackling this subject like no one else can.

Troubleshooting & Repair
320 pages • paperback • 8-3/8" x 10-7/8"
ISBN 0-7906-1224-0 • Sams 61224
$34.95

To order today or locate your nearest Prompt® Publications distributor at 1-800-428-7267 or www.samswebsite.com

Prices subject to change.

Servicing Digital Televisions

Robert Goodman

The growing popularity of digital televisions, as well as the increasing number of television stations broadcasting digitally, has brought this new technology into more homes than ever.

Yet little to no service data is available on digital televisions. Fear not, because *Servicing Digital Televisions* covers all aspects of this exciting new technology. Part of Sams Servicing Series, this comprehensive resource provides "tips and tricks" related to the equipment vs. specific models. *Servicing Digital Televisions* also goes into detail regarding popular models, with advice on how to apply the repair techniques to other manufacturers' sets.

Author Robert Goodman, CET, has devoted much of his career to developing and writing about more effective, efficient ways to troubleshoot electronics equipment. The author of more than 62 technical books and 150 technical articles, Goodman spends his time as a consultant and lecturer in Western Arkansas.

Troubleshooting & Repair
296 pages • paperback • 8-3/8" x 10-7/8"
ISBN 0-7906-1223-2 • Sams 61223
$34.95

CMOS Cookbook

Newton Braga

CMOS (Complementary Metal Oxide Semiconductors) ICs are an essential part of almost every electronics component and are not typically understood.

But no longer! Author Newton Braga takes the concepts from the legendary *CMOS Cookbook* from Don Lancaster (originally published by Sams/Macmillan) and brings them into the 21st century with this new and different look at CMOS technology.

Part of Sams Technical Publishing's Electronics Cookbook Series, *CMOS Sourcebook* is perfect for students as well as experienced designers. This comprehensive book covers CMOS technology from the top down—from theory to design applications.

Braga was the creator of the Brazilian edition of *Popular Electronics* magazine and is now technical director of two Latin American electronics magazines, *Revista Saber Electronica* and *Electronica Total*. Braga has authored more than 50 books in the United States and South America.

Electronics Technology
320 pages • paperback • 7 3/8 x 9 1/4"
ISBN 0-7906-1234-8 • Sams 61234
$39.95

**To order today or locate your nearest Prompt® Publications distributor
at 1-800-428-7267 or www.samswebsite.com**

Prices subject to change.

AGREEMENT

READ THIS AGREEMENT BEFORE OPENING THE SOFTWARE PACKAGE

BY OPENING THE SEALED PACKAGE YOU ACCEPT AND AGREE TO THE FOLLOWING TERMS AND CONDITIONS PRINTED BELOW. IF YOU DO NOT AGREE, DO NOT OPEN THE PACKAGE AND RETURN THE SEALED PACKAGE AND ALL MATERIALS YOU RECEIVED TO HOWARD W. SAMS & COMPANY, 2647 WATERFRONT PARKWAY EAST DRIVE SUITE 100 INDIANAPOLIS, IN 46214-2041 (HEREINAFTER "LICENSOR") WITHIN 30 DAYS OF RECEIPT ALONG WITH PROOF OF PAYMENT.

Licensor retains the ownership of this copy and any subsequent copies of the Software. This copy is licensed to you for use under the following conditions:

Permitted Uses. You may: use the Software on any supported computer configuration, provided the Software is sued on only one such computer and by one user at a time; permanently transfer the Software and its documentation to another user, provided you retain no copies and the recipient agrees to the terms of this Agreement.

Prohibited Uses. You may not: transfer, distribute, rent, sublicense, or lease the Software or documentation, except as provided herein; alter, modify, or adapt the Software or documentation, or portions thereof including, but not limited to, translation, decompiling, disassembling, or creating derivative works; make copies of the documentation, the Software, or portions thereof; export the Software.

LIMITED WARRANTY, DISCLAIMER OF WARRANTY

Licensor warrants that the optical media on which the Software is distributed is free from defects in materials and workmanship. Licensor will replace defective media at no charge, provided you return the defective media with dated proof of payment to Licensor within ninety (90) days of the date of receipt. This is your sole and exclusive remedy for any breach of warranty. EXCEPT AS SPECIFICALLY PROVIDED ABOVE, THE SOFTWARE IS PROVIDED ON AN "AS IS" BASIS. LICENSOR, THE AUTHOR, THE SOFTWARE DEVELOPERS, PROMPT PUBLICATIONS, HOWARD W. SAMS & COMPANY, AND BELL ATLANTIC MAKE NO WARRANTY OR REPRESENTATION, EITHER EXPRESS OR IMPLIED, WITH RESPECT TO THE SOFTWARE, INCLUDING ITS QUALITY, ACCURACY, PERFORMANCE, MERCHANTABILITY, OR FITNESS FOR A PARTICULAR PURPOSE. IN NO EVENT WILL LICENSOR, THE AUTHOR, THE SOFTWARE DEVELOPERS, PROMPT PUBLICATIONS, HOWARD W. SAMS & COMPANY, AND BELL ATLANTIC BE LIABLE FOR DIRECT, INDIRECT, SPECIAL, INCIDENTAL, OR CONSEQUENTIAL DAMAGES (INCLUDING BUT IS NOT LIMITED TO, INTERRUPTION OF SERVICE, LOSS OF DATA, LOSS OF CLASSROOM TIME, LOSS OF CONSULTING TIME) OR LOST PROFITS ARISING OUT OF THE USE OR INABILITY TO USE THE SOFTWARE OR DOCUMENTATION, EVEN IF ADVISED OF THE POSSIBILITY OF SUCH DAMAGES. IN NO CASE SHALL LIABILITY EXCEED THE AMOUNT OF THE FEE PAID. THE WARRANTY AND REMEDIES SET FORTH ABOVE ARE EXCLUSIVE AND IN LIEU OF ALL OTHERS, ORAL OR WRITTEN, EXPRESSED OR IMPLIED. Some states do not allow the exclusion or limitation of implied warranties or limitation of liability for incidental or consequential damages, so that the above limitation or exclusion may not apply to you.

GENERAL:

Licensor retains all rights, not expressly granted herein. This Software is copyrighted; nothing in this Agreement constitutes a waiver of Licensor's rights under United States copyright law. This License is nonexclusive. This License and your right to use the Software automatically terminate without notice from Licensor if you fail to Comply with any provision of this Agreement. This Agreement is governed by the laws of the State of Indiana.

LIBRARY
RUGBY COLLEGE